Fundamentals of Heat and Mass Transfer

Fundamentals of Heat and Mass Transfer

Mohan Gupta

RANDOM PUBLICATIONS
NEW DELHI (INDIA)

Fundamentals of Heat and Mass Transfer

ISBN 978-93-5111-410-9

Published in 2014 in India by

RANDOM PUBLICATIONS

4376-A/4B, Gali Murari Lal, Ansari Road
New Delhi-110 002
Phone : +91-11-43580356, +91-11-23289044
e-mail: randomexports@gmail.com, sales@randompublications.com, info@randompublications.com

Type Setting by : Keystoneprintads, Delhi-110051
Digitally Printed at : Replika Press Pvt. Ltd.

Preface

A communication network which does not depend on any physical connection needed to between two communication entities and have the flexibility to be mobile during communication. The current GSM and CDMA technology offers Mobile Communication. Mobile communication allows transmission of voice and multimedia data via a computer or a mobile device without having connected to any physical or fixed link. Mobile communication is evolving day by day and has become a must have for everyone. Mobile communication is the exchange of voice and data using a communication infrastructure at the same time regardless of any physical link.

Mobile Communications and Networking research is focused on next generation cellular wireless data network architectures. New technologies for improving spectral efficiency, spectrum utility and spatial reuse of spectrum on the radio-access network are being developed. On the mobile core network, research is oriented towards efficiently handling the exploding amount of mobile data traffic, enabling new revenue generating services for mobile operators and network CAPEX/OPEX reduction techniques. A variety of approaches including analysis, simulations, modeling, as well as experimental prototype design and network measurements are used to validate the technologies developed.

Mobile Technology has groomed a lot in the past few years, major reasons for rapid advancements in the mobile network technology is requirements for being mobile or connected to moves. Latest mobile handsets offer features which one had never thought off, ultimately it forces mobile network companies to bring these features in practice use to take commercial advantages.

Cellular is one of the fastest ongoing growing telecommunication industry in the world. It may easily be one of those industries which may never suffer from the economics of the world.

Wireless communications are a type of data communication that is performed and delivered wirelessly. This is a broad term that incorporates all procedures and forms of connecting and communicating between two or more devices using a wireless signal through wireless communication technologies and devices. Wireless communication is among technology's biggest contributions to mankind. Wireless communication involves the transmission of information over a distance without the help of wires, cables or any other forms of electrical conductors. The transmitted distance can be anywhere between a few meters (for example, a television's remote control) and thousands of kilometres (for example, radio communication).

The book is self-explanatory to a large extent and also serves as convenient reference for engineers working in the Mobile Communications field.

I thank all members of my team who have helped in the preparation of the book. My special thanks go to "Random Publications" who have published the book.

—Mohan Gupta

Contents

Preface .... *v-vi*

1. The Microscopic Interpretation of Heat and Work **1**

Quasi-static Processes 2
Feeling and Seeing Temperature Changes 9
Heat and Temperature 14
Specific Heats and Calorimetry 19

2. Combined Modes of Heat Transfer **25**

Introduction 25
Composite Bodies of Heat Transfer 27
Methods of Heat Transfer 29
Modes of Heat Transfer 37
Fourier Law of Heat Conduction 39
Newtion's Law of Cooling for Convection 40
Concept of Thermal Radiation 41
Electrical Circuits 44
Conduction Heat Transfer 46
Radiation Heat Transfer 48
Overall Heat Transfer Coefficient 52
Heat Flow through Wall with Variable Thermal Conductivity 56
Critical Thickness of Insulation 56

3. Mechanisms of Heat Transfer **60**

Heat Transfer in the Human Body 61
Fundamental Modes of Heat Transfer 63
Theory of Relativistic Heat Conduction 67
Convective Heat Transfer 72
Convection Cooling 74
Electromagnetic Radiation 77

Principle Mechanisms of Heat Transfer 81
Condensation Heat Transfer 86

4. Physical Concept of Entropy 93

Physical Significance of Entropy 93
Concept of Entropy 94
Ratio of the Entropy and Probability 102
The Conservation of Energy 107
Concept of Temperature 109
Heat 110
Complications with First Law 114
Significant and Remarkable Feature of Entropy 117
Change of Entropy of Gases 121
Entropy in Quantum Mechanics 128
Measurement of Entropy 132
Various Processe of Entropy Change 134
Entropy Change for Ideally Closed System Gas 137

5. Analogy between Fluid Friction and Heat Transfer 140

Introduction 140
Use of the Reynolds Analogy 143
Vertical Flate Plate in Free Convection Heat Transfer 145

6. Mass Transfer 149

Mixtures of Mass Transfer 150
Fick's Law of Diffusion 152
Definitions of Quantities 153

7. Fundamentals of Thermodynamics 155

Introduction 155
Principle of Thermodynamics 156
Process of Thermodynamic 161
Thermodynamic Systems 167
Thermodynamic Equilibrium 170
Engineering Thermodynamics 171
Thermodynamics Cycle 184

8. Mass of Carbon in Fuel Gases 188

Technologies for Measuring Combustion Flue Gases 190
The Analysis of Combustion Fuel Gas 195

9. Measurement of Heat Capacity 199

Concepts of Heat Capacity 200
Calculation from First Principles 201

Heat Capacity of Dimensionless 205
Hydrogen as a Fuel for Heat or Electricity 206

10. Specific Heats Energy 208

Heat Engines 210
Specific Heats of Solids 217
Specific Heats of Gases 225
Heat Capacity or Specific Heat 229

11. Enthalpy and Specific Heat Capacities of Ideal Gas 233

Definition of Gas 233
Perfect Gas Law 235
Enthalpy of Ideal Gas 235
P – V –T surface of an Ideal Gas 238
Triple Point 240
The Equation of State of an Ideal Gas 240

12. Heat Conduction 244

Introduction 244
General Heat Conduction Equation 245
The Controlling Coefficients 248
Magnitude of the Angular Velocity 250
Considering the One Dimensional Fin 251
One Dimensional Heat Conduction Equation 253
Transient Response of a Straight Fin 256
Heat Flow through Radial Systems 259
Heat Flow through a Plane Wall 263
Heat Transfer in Laminar Flow in Tubes 264
Heat Flow Conditions of Thermal Diffusivity 269
Initial and Boundary Conditions 270
The Heat Transfer of Thermal Energy 271
Thermal Conductivity of Perfect Infinite Crystals 275
Analysis of Heat Capacity 276
Internal Flow - Turbulent Flow Conditions 279

***Index* 284**

1

The Microscopic Interpretation of Heat and Work

Consider a macroscopic system A which is known to be in a given macrostate. To be more exact, consider an ensemble of similar macroscopic systems A, where each system in the ensemble is in one of the many microstates consistent with the given macrostate. There are two fundamentally different ways in which the average energy of A can change due to interaction with its surroundings. If the external parameters of the system remain constant then the interaction is termed a purely thermal interaction. Any change in the average energy of the system is attributed to an exchange of heat with its environment. Thus,

$$\Delta\bar{E} = Q,$$

where Q is the heat absorbed by the system. On a microscopic level, the energies of the individual microstates are unaffected by the absorption of heat. In fact, it is the distribution of the systems in the ensemble over the various microstates which is modified.

Suppose that the system A is thermally insulated from its environment. This can be achieved by surrounding it by an adiabatic envelope (*i.e.*, an envelope fabricated out of a material which is a poor conductor of heat, such a fibre glass). Incidentally, the term adiabatic is derived from the Greek adiabatos which means "impassable."

In scientific terminology, an adiabatic process is one in which there is no exchange of heat. The system A is still capable of interacting with its environment via its external parameters. This type of interaction is termed mechanical interaction, and any change in the average energy of the system is attributed to work done on it by its surroundings. Thus,

$$\Delta\bar{E} = -W,$$

where is the work done by the system on its environment. On a microscopic level, the energy of the system changes because the energies of the individual microstates are functions of the external parameters. Thus, if the external parameters are changed then, in general, the energies of all of the systems in the ensemble are modified (since each is in a specific microstate). Such a modification usually gives rise to a redistribution of the systems in the ensemble over the accessible microstates (without any heat exchange with the environment). Clearly, from a microscopic viewpoint, performing work on a macroscopic system is quite a complicated process. Nevertheless, macroscopic work is a quantity which can be readily measured experimentally. For instance, if the system A exerts a force F on its immediate surroundings, and the change in external parameters corresponds to a displacement x of the center of mass of the system, then the work done by A on its surroundings is simply,

$$W = F.x$$

i.e., the product of the force and the displacement along the line of action of the force. In a general interaction of the system A with its environment there is both heat exchange and work performed. We can write,

$$Q \equiv \Delta\bar{E} + W,$$

which serves as the general definition of the absorbed heat Q (hence, the equivalence sign). The quantity Q is simply the change in the mean energy of the system which is not due to the modification of the external parameters. Note that the notion of quantity of heat has no independent meaning apart from Eq. The mean energy $\bar{E}$ and work performed are both physical quantities which can be determined experimentally, whereas Q is merely a derived quantity.

QUASI-STATIC PROCESSES

A quasistatic process ensures that the system will go through a sequence of states that are infinitesimally close to equilibrium (so the system remains in quasistatic equilibrium), in which case the process is typically reversible. Any reversible process is necessarily a quasistatic one. However, some quasistatic processes are irreversible, if there is heat flowing (in or out of the system) or if entropy is being created in some other way. An example of a quasistatic process that is not reversible is a compression against a system with a piston subject to friction — Although the system is always in thermal equilibrium, the friction ensures the generation of dissipative entropy, which directly goes against the definition of reversible.

A notable example of a process that is not even quasistatic is the slow heat exchange between two bodies at two finitely different temperatures, where the heat exchange rate is controlled by an approximately adiabatic partition between the two bodies — in this case, no matter how slowly the process takes place, the states of the two bodies are never infinitesimally close to equilibrium, since thermal equilibrium requires that the two bodies be at precisely the same temperature. Some ambiguity exists in the literature concerning the distinction between quasistatic and reversible processes, as these are sometimes taken as synonyms. The reason is precisely because of the proven theorem that any reversible process is also a quasistatic one, even though we have also shown that the converse is not true.

It is practically not useful to differentiate between the two because any engineer would remember to include friction when calculating the dissipative entropy generation. The above definition is closer to the intuitive understanding of the word "quasi-" (almost) "static", while remaining technically different from reversible processes.

Consider the special case of an interaction of the system A with its surroundings which is carried out so slowly that A remains arbitrarily close to equilibrium at all times. Such a process is said to be quasi-static for the system A. In practice, a quasi-static process must be carried out on a time-scale which is much longer than the relaxation time of the system. Recall that the relaxation time is the typical time-scale for the system to return to equilibrium after being suddenly disturbed. A finite quasi-static change can be built up out of many infinitesimal changes. The infinitesimal heat $đW$ absorbed by the system when infinitesimal work $đW$ is done on its environment and its average energy changes by $d\bar{E}$ is given by,

$$đQ \equiv dE + đW.$$

The special symbols $đW$ and $đQ$ are introduced to emphasize that the work done and heat absorbed are infinitesimal quantities which do not correspond to the difference between two works or two heats. Instead, the work done and heat absorbed depend on the interaction process itself. Thus, it makes no sense to talk about the work in the system before and after the process, or the difference between these.

If the external parameters of the system have the values $x_1, \ldots, x_n$, then the energy of the system in a definite microstate can be written,

$$E_r = E_r(x_1, \cdots, x_n).$$

Hence, if the external parameters are changed by infinitesimal amounts, so that x_a ® $x_a + dx_a$ for in the range 1 to , then the corresponding change in the energy of the microstate is,

$$dE_r = \sum_{a=1}^{n} \frac{\partial E_r}{\partial x_\alpha} dx_\alpha.$$

The work $đW$ done by the system when it remains in this particular state r is,

$$đW_r = -dE_r = \sum_{\alpha=1}^{n} X_{\alpha r} dx_\alpha,$$

where,

$$X_{\alpha r} \equiv -\frac{\partial E_r}{\partial x_\alpha}$$

is termed the generalized force (conjugate to the external parameter x_a) in the state r. Note that if is x_a a displacement then X_{ar} is an ordinary force. Consider now an ensemble of systems. Provided that the external parameters of the system are changed quasi-statically, the generalized forces X_{ar} have well defined mean values which are calculable from the distribution of systems in the ensemble characteristic of the instantaneous macrostate. The macroscopic work $đW$ resulting from an infinitesimal quasi-static change of the external parameters is obtained by calculating the decrease in the mean energy resulting from the parameter change. Thus,

$$đW = \sum_{\alpha=1}^{n} \bar{X}_\alpha dx_\alpha,$$

where,

$$\bar{X}_\alpha \equiv -\overline{\frac{\partial E_r}{\partial x_\alpha}}$$

is the mean generalized force conjugate to x_a. The mean value is calculated from the equilibrium distribution of systems in the ensemble corresponding to the external parameter values x_a. The macroscopic work W resulting from a finite quasi-static change of external parameters can be obtained by integrating equation.

The most well-known example of quasi-static work in thermodynamics is that done by pressure when the volume changes. For simplicity, suppose that the volume V is the only external parameter of any consequence. The work done in changing the volume from V to $V + dV$ is simply the product of the force and the displacement (along the line of action of the force). By definition, the mean equilibrium pressure $\bar{p}$ of a given macrostate is equal to the normal force per unit area acting on any surface element. Thus, the normal force acting on a surface element dS_i is $\bar{p}\, dS_i$. Suppose that the surface element is subject to a displacement dx_i. The work done by the element is $\bar{p}\, dS_i.dx_i$. The total work done by the system is obtained by summing over all of the surface elements. Thus,

$$đW = \bar{p}\, dV,$$

where,

$$dV = \sum_i dS_i.dx_i$$

is the infinitesimal volume change due to the displacement of the surface. It follows from that,

$$\bar{p} = -\frac{\partial \bar{E}}{\partial V},$$

so the mean pressure is the generalized force conjugate to the volume V. Suppose that a quasi-static process is carried out in which the volume is changed from V_i to V_f. In general, the mean pressure is a function of the volume, so $\bar{p} = \bar{p}$ (V). It follows that the macroscopic work done by the system is given by,

$$W_{if} = \int_{V_i}^{V_f} \text{đ}W = \int_{V_i}^{V_f} \bar{p}(V)\, dV.$$

This quantity is just the "area under the curve" in a plot of $\bar{p}$ (V) versus V.

EXACT AND INEXACT DIFFERENTIALS

In our investigation of heat and work we have come across various infinitesimal objects such as $d\bar{E}$ and $\text{đ}W$. It is instructive to examine these infinitesimals more closely. Consider the purely mathematical problem where $F(x, y)$is some general function of two independent variables x and y. Consider the change in F in going from the point (x, y) in the x–y plane to the neighbouring point $(x + dx, y + dy)$. This is given by

$$dF = F(x + dx, y + dy) - F(x, y),$$

which can also be written,

$$dF = X(x, y)dx + Y(x, y)dy,$$

where X = ¶F/¶x and Y = ¶F/¶y. Clearly, dF is simply the infinitesimal difference between two adjacent values of the function F. This type of infinitesimal quantity is termed an exact differential to distinguish it from another type to be discussed presently. If we move in the x–y plane from an initial point i º (x_i, y_i) to a final point f º (x_f, y_f) then the corresponding change in is F given by,

$$\Delta F = F_f - F_i = \int_i^f dF = \int_i^f (X\, dx + Y\, dy).$$

Note that since the difference on the left-hand side depends only on the initial and final points, the integral on the right-hand side can only depend on these points as well. In other words, the value of the integral is independent of the path taken in going from the initial to the final point. This is the distinguishing feature of an exact differential.

Consider an integral taken around a closed circuit in the $x - y$ plane. In this case, the initial and final points correspond to the same point, so the difference $F_f - F_i$ is clearly zero. It follows that the integral of an exact differential over a closed circuit is always zero:

$$\oint dF \equiv 0.$$

Of course, not every infinitesimal quantity is an exact differential. Consider the infinitesimal object,

$$đG \equiv X'(x, y)dx + Y'(x, y)dz,$$

where X¢ and Y¢re two general functions of x and y. It is easy to test whether or not an infinitesimal quantity is an exact differential. Consider the expression. It is clear that since X = ¶F/¶x and Y = ¶F/¶y then,

$$\frac{\partial X}{\partial y} = \frac{\partial Y}{\partial x} = \frac{\partial^2 F}{\partial x \partial y}.$$

Thus, if,

$$\frac{\partial X'}{\partial y} \neq \frac{\partial Y'}{\partial x}$$

(as is assumed to be the case), then $đG$ cannot be an exact differential, and is instead termed an inexact differential. The special symbol $đ$ is used to denote an inexact differential. Consider the integral of $đG$ over some path in the $x - y$ plane. In general, it is not true that,

$$\int_i^f đG = \int_i^f (X'dx + Y'dy)$$

is independent of the path taken between the initial and final points. This is the distinguishing feature of an inexact differential. In particular, the integral of an inexact differential around a closed circuit is not necessarily zero, so,

$$\oint đG \neq 0.$$

Consider, for the moment, the solution of,

$$đG = 0,$$

which reduces to the ordinary differential equation,

$$\frac{dy}{dx} = -\frac{X'}{Y'}.$$

Since the right-hand side is a known function of x and y, the above equation defines a definite direction (*i.e.*, gradient) at each point in the x–y plane. The solution simply consists of drawing a system of curves in the x–y plane such that at any point the tangent to the curve is as specified in equation. This defines a set of curves which can be written s(x, y) = c, where c is a labeling parameter. It follows that,

$$\frac{d\sigma}{dx} \equiv \frac{\partial \sigma}{\partial x} + \frac{\partial \sigma}{\partial y}\frac{dy}{dx} = 0.$$

The elimination of dy/dx between Eqs. yields,

$$Y'\frac{\partial \sigma}{\partial x} = X'\frac{\partial \sigma}{\partial y} = \frac{X'Y'}{\tau},$$

where t(x, y)is function of x and y. The above equation could equally well be written,

$$X' = \frac{\partial \sigma}{\partial x}, \quad Y' = \tau\frac{\partial \sigma}{\partial y}.$$

Inserting equation gives,

$$đG = \tau\left(\frac{\partial \sigma}{\partial x}dx + \frac{\partial \sigma}{\partial y}dy\right) = \tau\, d\sigma,$$

or,

$$\frac{đG}{\tau} = d\sigma.$$

Thus, dividing the inexact differential $đG$ by t yields the exact differential ds. A factor t which possesses this property is termed an integrating factor.

Since the above analysis is quite general, it is clear that an inexact differential involving two independent variables always admits of an integrating factor. Note, however, this is not generally the case for inexact differentials involving more than two variables.

After this mathematical excursion, let us return to physical situation of interest. The macrostate of a macroscopic system can be specified by the values of the external parameters (*e.g.*, the volume) and the mean energy $\overline{E}$. This, in turn, fixes other parameters such as the mean pressure $\overline{p}$. Alternatively, we can specify the external parameters and the mean pressure, which fixes the mean energy. Quantities such as $d\overline{p}$ and $d\overline{E}$ are infinitesimal differences between well-defined quantities: *i.e.*, they are exact differentials.

For example, $d\overline{E} = \overline{E}_f - \overline{E}_i$ is just the difference between the mean energy of the system in the final macrostate f and the initial macrostate i, in the limit where these two states are nearly the same. It follows that if the system is taken from an initial macrostate i to any final macrostate f the mean energy change is given by,

$$\Delta\overline{E} = \overline{E}_f - \overline{E}_i = \int_i^f d\overline{E}.$$

However, since the mean energy is just a function of the macrostate under consideration, $\bar{E}_f$ and $\bar{E}_i$ depend only on the initial and final states, respectively. Thus, the integral $\int d\bar{E}$ depends only on the initial and final states, and not on the particular process used to get between them.

Consider, now, the infinitesimal work done by the system in going from some initial macrostate to some neighbouring final macrostate f. In general, $đW = \sum \bar{X}_\alpha \bar{d}x_\alpha$ is not the difference between two numbers referring to the properties of two neighbouring macrostates. Instead, it is merely an infinitesimal quantity characteristic of the process of going from state to state f. In other words, the work $đW$ is in general an inexact differential. The total work done by the system in going from any macrostate to some other macrostate f can be written as,

$$W_{if} = \int_i^f đW_1$$

where the integral represents the sum of the infinitesimal amounts of work $đW$ performed at each stage of the process. In general, the value of the integral does depend on the particular process used in going from macrostate to macrostate f.

Recall that in going from macrostate i to macrostate f the change $\Delta\bar{E}$ does not depend on the process used whereas the work W, in general, does. Thus, it follows from the first law of thermodynamics, equation that the heat Q, in general, also depends on the process used. It follows that,

$$đQ \equiv d\bar{E} + đW$$

is an inexact differential. However, by analogy with the mathematical example, there must exist some integrating factor, T, say, which converts the inexact differential $đQ$ into an exact differential. So,

$$\frac{đQ}{T} \equiv dS.$$

It will be interesting to find out what physical quantities correspond to the functions T and S.

Suppose that the system is thermally insulated, so that $Q = 0$. In this case, the first law of thermodynamics implies that ,

$$W_{if} = -\Delta\bar{E}.$$

Thus, in this special case, the work done depends only on the energy difference between in the initial and final states, and is independent of the process. In fact, when Clausius first formulated the first law in 1850 this is how he expressed it:

If a thermally isolated system is brought from some initial to some final state then the work done by the system is independent of the process used.

If the external parameters of the system are kept fixed, so that no work is done, then $đW = 0$, equation reduces to,

$$đQ - d\bar{E},$$

and $đQ$ becomes an exact differential. The amount of heat Q absorbed in going from one macrostate to another depends only on the mean energy difference between them, and is independent of the process used to effect the change. In this situation, heat is a conserved quantity, and acts very much like the invisible indestructible fluid of Lavoisier's calorific theory.

FEELING AND SEEING TEMPERATURE CHANGES

According to new research, people exposed to warmer temperatures find it harder to grasp viewpoints other than their own, while those exposed to colder ones find it easier: to take the heat out of a disagreement, it seems, you should literally take the heat out of the room. Since I've always preferred the cold, – I'm already sick of people hoping this summer will be a hot one – this was music (or air conditioning?) to my ears. It's tempting to extrapolate: might this explain the affable tolerance of Canadians, say, or the history of prejudice in the US south? Sadly, on closer reading, the study is only a partial victory for cold. We're better at seeing other perspectives when we're chilly, the researchers argue, because cold triggers a sense of social distance. It reminds us of our separateness, and thus the fact that others aren't like us. We gain perspective at the cost of intimacy.

So what looks, at first, like a surprising result turns out to reinforce one of the most intriguing psychological findings of recent years: that coldness makes people feel lonely. The opposite's also true: loneliness makes people feel cold. In one experiment, students played a computer game in which they threw a ball back and forth with other on-screen characters, each of whom they (wrongly) believed was controlled by another student, playing elsewhere. After a while, the others sometimes began to keep the ball to themselves. Subsequently, players who'd been thus ostracised showed a marked preference for hot foods over cold ones; non-ostracised players didn't. In a recent rerun of the experiment, ostracism led to a drop in skin temperature. Other studies have found that hot baths relieve loneliness, and that merely being reminded of an experience of exclusion prompts people to judge a room's temperature as colder.

This kind of research – about how seemingly innocuous aspects of our surroundings can exert powerful effects – has been in the doghouse lately;

several classic findings have proved difficult to replicate. It's no longer clear, for example, whether being exposed to words associated with old age ("grey", "bingo") really does make people start walking more slowly. But there's reason to believe the link between loneliness and temperature will hold up. It's no mere matter of word association: temperature may be a crucial way our bodies keep track of whether we're getting the social contact we need. It's easy to see why natural selection might have given us a yearning to be near friendly fellow tribe members: they were crucial for food, security and sex. People worry that social media are making us lonely and isolated, but what if that is exactly half-true? What if they are not making us isolated – online connections are real, after all – but are making us feel lonely, partly because those connections don't involve heat?

Within some reasonable temperature range, we can get a rough idea how warm something is by touching it. But this can be unreliable—if you put one hand in cold water, one in hot, then plunge both of them into lukewarm water, one hand will tell you it's hot, the other will feel cold. For something too hot to touch, we can often get an impression of how hot it is by approaching and sensing the radiant heat. If the temperature increases enough, it begins to glow.

The problem with these subjective perceptions of heat is that they may not be the same for everybody. If our two hands can't agree on whether water is warm or cold, how likely is it that a group of people can set a uniform standard? We need to construct a device of some kind that responds to temperature in a simple, measurable way—we need a thermometer. The first step on the road to a thermometer was taken by one Philo of Byzantium, an engineer, in the second century BC. He took a hollow lead sphere connected with a tight seal to one end of a pipe, the other end of the pipe being under water in another vessel.

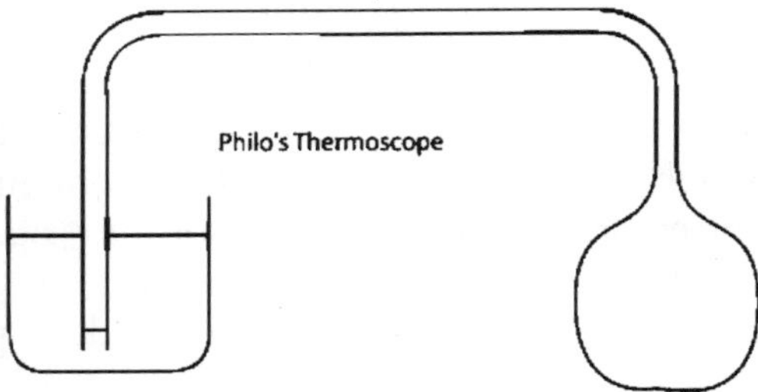

Fig. Philo's Thermoscope

To quote Philo: "...if you expose the sphere to the sun, part of the air enclosed in the tube will pass out when the sphere becomes hot. This will be evident because the air will descend from the tube into the water, agitating it and producing a succession of bubbles. Now if the sphere is put back in the shade, that is, where the sun's rays do not reach it, the water will rise and pass through the tube ..." "No matter how many times you repeat the operation, the same thing will happen.

In fact, if you heat the sphere with fire, or even if you pour hot water over it, the result will be the same." Notice that Philo did what a real investigative scientist should do—he checked that the experiment was reproducible, and he established that the air's expansion was in response to heat being applied to the sphere, and was independent of the source of the heat.

This expansion of air on heating became widely known in classical times, and was used in various dramatic devices. For example, Hero of Alexandria describes a small temple where a fire on the altar causes the doors to open. The altar is a large airtight box, with a pipe leading from it to another enclosed container filled with water. When the fire is set on top of the altar, the air in the box heats up and expands into a second container which is filled with water. This water is forced out through an overflow pipe into a bucket hung on a rope attached to the door hinges in such a way that as the bucket fills with water, it drops, turns the hinges, and opens the doors.

The pipe into this bucket reaches almost to the bottom, so that when the altar fire goes out, the water is sucked back and the doors close again. Still, none of these ingenious devices is a thermometer.

There was no attempt by Philo or his followers to make a quantitative measurement of how hot or cold the sphere was. And the "meter" in thermometer means measurement.

THE FIRST THERMOMETER

Galileo claimed to have invented the first thermometer. Well, actually, he called it a thermoscope, but he did try to measure "degrees of heat and cold" according to a colleague, and that qualifies it as a thermometer. Galileo used an inverted narrow-necked bulb with a tubular neck, like a hen's egg with a long glass tube attached at the tip.

He first heated the bulb with his hands then immediately put it into water. He recorded that the water rose in the bulb the height of "one palm". Later, either Galileo or his colleague

Santorio Santorio put a paper scale next to the tube to read off changes in the water level. This definitely made it a thermometer, but who thought of it first isn't clear (they argued about it). And, in fact, this thermometer had problems.

Question: what problems? If you occasionally top up the water, why shouldn't this thermometer be good for recording daily changes in temperature?

Answer: because it's also a barometer! But—Galileo didn't know about the atmospheric pressure.

Torricelli, one of Galileo's pupils, was the first to realise, shortly after Galileo died, that the real driving force in suction was external atmospheric pressure, a satisfying mechanical explanation in contrast to the philosophical "nature abhors a vacuum". In the 1640's, Pascal pointed out that the variability of atmospheric pressure rendered the air thermometer untrustworthy. Liquid-in-glass thermometers were used from the 1630's, and they were of course insensitive to barometric pressure. Meteorological records were kept from this time, but there was no real uniformity of temperature measurement until Fahrenheit, almost a hundred years later.

NEWTON'S ANONYMOUS TABLE OF TEMPERATURES

Presumably he felt that this project lacked the timeless significance of some of his other achievements. Taking the freezing point of water as zero, Newton found the temperature of boiling water to be almost three times that of the human body, melting lead eight times as great (actually 327 °C, whereas 8 × 37 = 296, so this is pretty good!) but for higher temperatures, such as that of a wood fire, he underestimated considerably. He used a linseed oil liquid in glass thermometer up to the melting point of tin (232 °C). Newton tried to estimate the higher temperatures indirectly. He heated up a piece of iron in a fire, then let it cool in a steady breeze.

He found that, at least at the lower temperatures where he could cross check with his thermometer, the temperature dropped in a geometric progression, that is, if it took five minutes to drop from 80° above air temperature to 40° above air temperature, it took another five minutes to drop to 20° above air, another five to drop to 10° above, and so on. He then assumed this same pattern of temperature drop was true at the high temperatures beyond the reach of his thermometer, and so estimated the temperature of the fire and of iron glowing red hot. This wasn't very accurate—he (under)estimated the temperature of the fire to be about 600 °C.

FAHRENHEIT'S EXCELLENT THERMOMETER

The first really good thermometer, using mercury expanding from a bulb into a capillary tube, was made by Fahrenheit in the early 1720's. He got the idea of using mercury from a colleague's comment that one should correct a barometer reading to allow for the variation of the density of mercury with temperature.

The point that has to be borne in mind in constructing thermometers, and defining temperature scales, is that not all liquids expand at uniform rates on heating—water, for example, at first contracts on heating from its freezing point, then begins to expand at around forty degrees Fahrenheit, so a water thermometer wouldn't be very helpful on a cold day. It is also not easy to manufacture a uniform cross section capillary tube, but Fahrenheit managed to do it, and demonstrated his success by showing his thermometers agreed with each other over a whole range of temperatures. Fortunately, it turns out that mercury is well behaved in that the temperature scale defined by taking its expansion to be uniform coincides very closely with the true temperature scale.

Amontons' Air Thermometer

A little earlier (1702) Amontons introduced an air pressure thermometer. He established that if air at atmospheric pressure at the freezing point of water is enclosed then heated to the boiling point of water, but meanwhile kept at constant volume by increasing the pressure on it, the pressure goes up by about 10 inches of mercury.

He also discovered that if he compressed the air in the first place, so that it was at a pressure of sixty inches of mercury at the temperature of melting ice, then if he raised its temperature to that of boiling water, at the same time adding mercury to the column to keep the volume of air constant, the pressure increased by 20 inches of mercury. In other words, he found that for a fixed amount of air kept in a container at constant volume, the pressure increased with temperature by about 33 per cent from freezing to boiling, that percentage being independent of the initial pressure.

THE ZEROTH LAW OF THERMODYNAMICS

Once the thermometer came to be widely used, more precise observations of temperature and heat flow became possible. Joseph Black, a professor at the University of Edinburgh in the 1700's, noticed that a collection of objects at different temperatures, if brought together, will all eventually reach the same temperature.

"By the use of these instruments [thermometers] we have learned, that if we take 1000, or more, different kinds of matter, such as metals, stones, salts, woods, cork, feathers, wool, water and a variety of other fluids, although they be all at first of different heats, let them be placed together in a room without a fire, and into which the sun does not shine, the heat will be communicated from the hotter of these bodies to the colder, during some hours, perhaps, or the course of a day, at the end of which time, if we apply a thermometer to all of them in succession, it will point to precisely the same degree."

We say nowadays that bodies in "thermal contact" eventually come into "thermal equilibrium"—which means they finally attain the same temperature, after which no further heat flow takes place. This is equivalent to:

If two objects are in thermal equilibrium with a third, then they are in thermal equilibrium with each other.

The "third body" in a practical situation is just the thermometer. It's perhaps worth pointing out that this trivial sounding statement certainly wasn't obvious before the invention of the thermometer. With only the sense of touch to go on, few people would agree that a piece of wool and a bar of metal, both at 0°C, were at the same temperature. "flow of heat" that takes place between bodies as they move towards thermal equilibrium? For example, suppose I reproduce one of Fahrenheit's experiments, by taking 100 ccs of water at 100 °F, and 100ccs at 150 °F, and mix them together in an insulated jug so little heat escapes. What is the final temperature of the mix? Of course, it's close to 125°F—not surprising, but it does tell us something! It tells us that the amount of heat required to raise the temperature of 100 cc of water from 100 °F to 125 °F is exactly the same as the amount needed to raise it from 125 °F to 150 °F.

A series of such experiments established that it always took the same amount of heat to raise the temperature of 1 cc of water by one degree, independent of the initial temperature of the water (provided it's between the freezing point and the boiling point). This makes it possible to define a unit of heat. Perhaps unfairly to Fahrenheit, 1 calorie is the heat required to raise the temperature of 1 gram of water by 1 degree Celsius.

(Celsius also lived in the early 1700's. His scale has the freezing point of water as 0 °C, the boiling point as 100°C. Fahrenheit's scale is no longer used in science, but lives on in engineering in the US, and in the British Thermal Unit, which is the heat required to raise the temperature of one pound of water by 1 °F).

HEAT AND TEMPERATURE

Heat, energy and temperature pervade our lives. Just think about it. We give concentration to hot and cold in deciding what we wear during the day, at night and when we go to bed. We think about the topic when deciding how many covers, if any at all, we will wrap in or sleep under at night in order to maintain *the right temperature*. Many of us have heating and cooling systems in our homes, schools and work places that control the temperature during the day and night to keep us as comfortable as possible without spending too much money.

We install fans or use portable fans in our homes to keep us comfortable. Most of us have cars equipped with heating and air conditioning systems; some may even have meters in their cars that register the indoor and outdoor temperatures. Many of us watch and listen to weather reports, especially the forecasted temperatures, with great interest so that we can make decisions about what to wear and what to do on the following day.

Our bodies are highly sensitive to hot and cold. We learn very early in life through the University of Pain that we shouldn't touch a hot pot on the stove or a hot light bulb in a lamp. It is a remarkable lesson that makes an impression for a lifetime. In that same university, we learn that we should be careful about mouthing or tasting hot foods.

We learn how to use our hands to feel the heat that emanates from such foods and we learn how to blow gently on the food to help cool it down. Those of us with either poor genes or poor dental care (or both) know the pain of cold ice cream hitting a nerve in a tooth cavity. And we all have vivid memories of Mom and Dad sticking a thermometer under our tongue (or elsewhere) to get our body temperature to see if we had a fever. We all know the experience of perspiring-our bodies built-in mechanism of cooling us down when temperatures begin to rise. Our bodies have a narrow range of temperatures that they can maintain. Any departure from this range can result in major consequences, including death. Energy topics have become topics of national and global interests. Politicians and scholars debate topics associated with energy supplies, energy alternatives and the impact of our energy dependencies upon the environment.

Global *warming* is a *hot topic* in both national and global circles. As the price of octane gasoline rises at the pump, our interest in energy efficient transportation becomes heightened. Scientists search passionately for alternative fuels which will be cost effective and environmentally friendly. We often hear of an energy crisis or even an energy shortage. Meanwhile, scientists preach about the law of *conservation of energy*, leaving the public confused about how there can be a shortage of something that is *conserved*.

TEMPERATURE

We all have *a feel* for what temperature is. We even have a shared language that we use to qualitatively describe temperature. The water in the shower or bathtub feels hot or cold or warm. The weather outside is *chilly* or *steamy*. We certainly have a good feel for how one temperature is qualitatively different than another temperature. We may not always agree on whether the room temperature is too hot or too cold or just right. But we will likely all agree that we possess built-in thermometers for making qualitative judgments about relative temperatures.

How a Thermometer Works

Today, there are a variety of types of thermometers. The type that most of us are familiar with from science class is the type that consists of a liquid encased in a narrow glass column. Older thermometers of this type used liquid mercury. In response to our understanding of the health concerns associated with mercury exposure, these types of thermometers usually use some type of liquid alcohol. These liquid thermometers are based on the principal of thermal expansion. When a substance gets hotter, it expands to a greater volume. Nearly all substances exhibit this behaviour of thermal expansion. It is the basis of the design and operation of thermometers.

As the temperature of the liquid in a thermometer increases, its volume increases. The liquid is enclosed in a tall, narrow glass (or plastic) column with a constant cross-sectional area. The increase in volume is thus due to a change in height of the liquid within the column.

The increase in volume, and thus in the height of the liquid column, is proportional to the increase in temperature. Suppose that a 10-degree increase in temperature causes a 1-cm increase in the column's height. Then a 20-degree increase in temperature will cause a 2-cm increase in the column's height. And a 30-degree increase in temperature will causes 3-cm increase in the column's height. The relationship between the temperature and the column's height is linear over the small temperature range for which the thermometer is used. This linear relationship makes the calibration of a thermometer a relatively easy task.

The calibration of any measuring tool involves the placement of divisions or marks upon the tool to measure a quantity accurately in comparison to known standards. Any measuring tool-even a meter stick-must be calibrated. The tool needs divisions or markings; for instance, a meter stick typically has markings every 1-cm apart or every 1-mm apart. These markings must be accurately placed and the accuracy of their placement can only be judged when comparing it to another object known to have an accurate length.

A thermometer is calibrated by using two objects of known temperatures. The typical process involves using the freezing point and the boiling point of water. Water is known to freeze at 0°C and to boil at 100°C at an atmospheric pressure of 1 atm. By placing a thermometer in mixture of ice water and allowing the thermometer liquid to reach a stable height, the 0-degree mark can be placed upon the thermometer. Similarly, by placing the thermometer in boiling water (at 1 atm of pressure) and allowing the liquid level to reach a stable height, the 100-degree mark can be placed upon the thermometer. With these two markings placed upon the thermometer, 100 equally spaced divisions can be placed between them to represent the 1-degree marks. Since there is a linear relationship between the temperature and the height of the liquid, the divisions between 0 degree and 100 degree can be equally spaced. With a calibrated thermometer, accurate measurements can be made of the temperature of any object within the temperature range for which it has been calibrated.

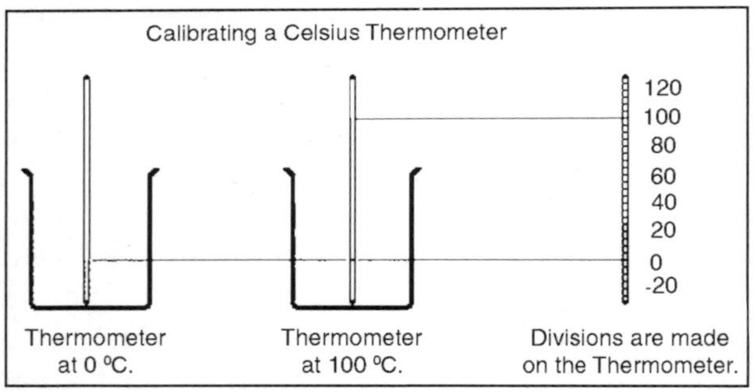

Temperature Scales

The thermometer calibration process results in what is known as a centigrade thermometer. A centigrade thermometer has 100 divisions or intervals between the normal freezing point and the normal boiling point of water. Today, the centigrade scale is known as the Celsius scale, named after the Swedish astronomer Anders Celsius who is credited with its development. The Celsius scale is the most widely accepted temperature scale used throughout the world. It is the standard unit of temperature measurement in nearly all countries, the most notable exception being the United States. Using this scale, a temperature of 28 degrees Celsius is abbreviated as 28°C. Usually slow to adopt the metric system and other accepted units of measurements, the United States more commonly uses the Fahrenheit temperature scale.

A thermometer can be calibrated using the Fahrenheit scale in a similar manner. The difference is that the normal freezing point of water is designated as 32 degrees and the normal boiling point of water is designated as 212 degrees in the Fahrenheit scale. As such, there are 180 divisions or intervals between these two temperatures when using the Fahrenheit scale. The Fahrenheit scale is named in honour of German physicist Daniel Fahrenheit. A temperature of 76 degree Fahrenheit is abbreviated as 76°F. In most countries throughout the world, the Fahrenheit scale has been replaced by the use of the Celsius scale.

Temperatures expressed by the Fahrenheit scale can be converted to the Celsius scale equivalent using the equation below:

$$°C = (°F - 32°)/1.8$$

Similarly, temperatures expressed by the Celsius scale can be converted to the Fahrenheit scale equivalent using the equation below:

$$°F = 1.8 \bullet °C + 32°$$

The Kelvin Temperature Scale

The Celsius and Fahrenheit scales are the most widely used temperature scales, there are several other scales that have been used throughout history. For example, there is the Rankine scale, the Newton scale and the Romer scale, all of which are rarely used.

Finally, there is the Kelvin temperature scale, which is the standard metric system of temperature measurement and perhaps the most widely used temperature scale used among scientists. The Kelvin temperature scale is similar to the Celsius temperature scale in the sense that there are 100 equal degree increments between the normal freezing point and the normal boiling point of water.

However, the zero-degree mark on the Kelvin temperature scale is 273.15 units cooler than it is on the Celsius scale. So a temperature of 0 Kelvin is equivalent to a temperature of –273.15 °C. Observe that the degree symbol is not used with this system. So a temperature of 300 units above 0 Kelvin is referred to as 300 Kelvin and not 300 degree Kelvin; such a temperature is abbreviated as 300 K.

Conversions between Celsius temperatures and Kelvin temperatures (and *vice versa*) can be performed using one of the two equations below.

$$°C = K - 273.15°$$

$$K = °C + 273.15$$

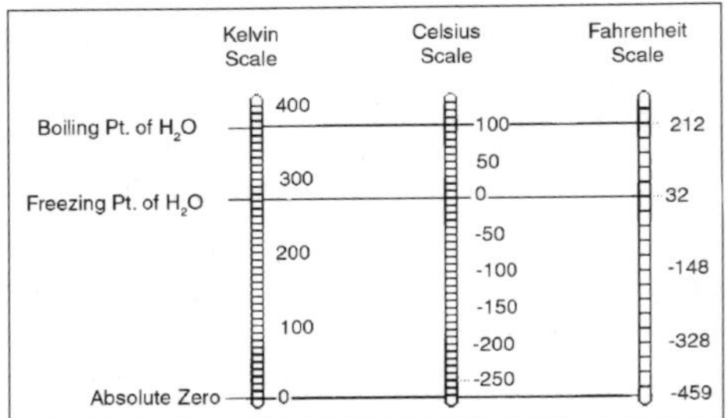

The zero point on the Kelvin scale is known as absolute zero. It is the lowest temperature that can be achieved. The concept of an absolute temperature minimum was promoted by Scottish physicist William Thomson (a.k.a. Lord Kelvin) in 1848. Thomson theorized based on thermodynamic principles that the lowest temperature which could be achieved was –273°C. Prior to Thomson, experimentalists such as Robert Boyle (late 17th century) were well aware of the observation that the volume (and even the pressure) of a sample of gas was dependent upon its temperature.

Measurements of the variations of pressure and volume with changes in the temperature could be made and plotted. Plots of volume vs. temperature (at constant pressure) and pressure vs. temperature (at constant volume) reflected the same conclusion-the volume and the pressure of a gas reduces to zero at a temperature of –273°C.

Since these are the lowest values of volume and pressure that are possible, it is reasonable to conclude that –273°C was the lowest temperature that was possible.

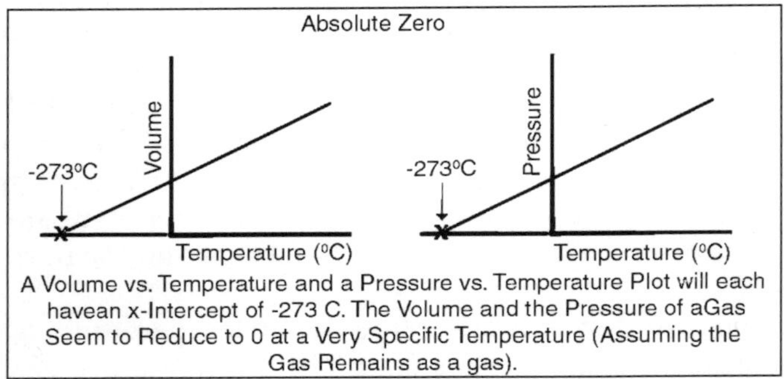

A Volume vs. Temperature and a Pressure vs. Temperature Plot will each havean x-Intercept of -273 C. The Volume and the Pressure of aGas Seem to Reduce to 0 at a Very Specific Temperature (Assuming the Gas Remains as a gas).

Thomson referred to this minimum lowest temperature as *absolute zero* and argued that a temperature scale be adopted that had absolute zero as the lowest value on the scale. Today, that temperature scale bears his name. Scientists and engineers have been able to cool matter down to temperatures close to -273.15°C, but never below it. In the process of cooling matter to temperatures close to absolute zero, a variety of unusual properties have been observed. These properties include superconductivity, superfluidity and a state of matter known as a Bose-Einstein condensate.

SPECIFIC HEATS AND CALORIMETRY

The specific heat of a substance is the heat required in calories to raise the temperature of 1 gram by 1 degree Celsius. As Fahrenheit continues his measurements of heat flow, it quickly became evident that for different materials, the amount of heat needed to raise the temperature of one gram by one degree could be quite different. It was natural to measure specific heats relative to that of water, the simplest and most readily available substance for calorimetric experiments.

There were some surprises. For example, it had been widely thought before the measurements were made, that one cc of Mercury, being a lot heavier than one cc of water, would take more heat to raise its temperature by one degree. This proved not to be the case—Fahrenheit himself made the measurement. In an insulating container, called a "calorimeter" he added 100 ccs of water at 100 °F to 100 ccs of mercury at 150 °F, and stirred so they quickly reached thermal equilibrium.

Question: What do you think the final temperature was?

Answer: The final temperature was, surprisingly, about 120 °F. 100 cc of water evidently "contained more heat" than 100 cc of mercury, despite the large difference in weight.

The technique, called calorimetry, was widely used to find specific heats of many different substances, and at first no clear pattern emerged. It was puzzling that the specific heat of mercury was so low compared with water. As the experiments progressed, it gradually became evident that heavier substances, paradoxically, had lower specific heats.

Meanwhile, this quantitative approach to scientific observation had spread to chemistry. Towards the end of the 1700's, Lavoisier weighed chemicals involved in reactions before and after the reaction.

This involved weighing the gases involved, so had to be carried out in closed containers, so that, for example, the weight of oxygen used and the carbon dioxide, etc., produced would accounted for in studying combustion. The big discovery was that mass was neither created nor destroyed. This had not been realised before because no one had weighed the gases involved. It made the atomic theory suddenly more plausible, with the idea that maybe chemical reactions were just rearrangements of atoms into different combinations.

Lavoisier also clarified the concept of an element, an idea that was taken up in about 1800 by John Dalton, who argues that a given compound consisted of identical molecules, made up of elementary atoms in the same proportion, such as H_2O (although that was thought initially to be HO). This explained why, when substances reacted chemically, such as the burning of hydrogen to form water, it took exactly eight grams of oxygen for each gram of hydrogen. (produce H_2O_2 under the right conditions, with exactly sixteen grams of oxygen to one of hydrogen, but the simple ratios of amounts of oxygen needed for the two reactions were simply explained by different molecular structures, and made the atomic hypothesis even more plausible.)

Much effort was expended carefully weighing the constituents in many chemical reactions, and constructing diagrams of the molecules. The important result of all this work was that it became possible to list the relative weights of the atoms involved. For example, the data on H_2O and H_2O_2 led to the conclusion that an oxygen atom weighed sixteen times the weight of a hydrogen atom. It must be emphasized, though, that these results gave no clue as to the actual weights of atoms! All that was known was that atoms were too small to see in the best microscopes. Nevertheless, knowing the relative weights of some atoms in 1820 led to an important discovery.

Two professors in France, Dulong and Petit, found that for a whole series of elements the product of atomic weight and specific heat was the same.

Element	Specific Heat (cals/gm/ degree C)	Relative weights of the atoms	Product of relative atomic weight and specific heat
Lead	0.0293	12.95	0.3794
Tin	0.0514	7.35	0.3779
Zinc	0.0927	4.03	0.3736
Sulphur	0.188	2.011	0.378

The significance of this, as they pointed out, was that the "specific heat", or heat capacity, of each atom was the same—a piece of lead and a piece of zinc having the same number of atoms would have the same heat capacity. So heavier atoms absorbed no more heat than lighter atoms for a given rise in temperature.

This partially explained why mercury had such a surprisingly low heat capacity. Of course, having no idea how big the atoms might be, they could go no further. And, indeed, many of their colleagues didn't believe in atoms anyway, so it was hard to convince them of the significance of this discovery.

LATENT HEAT

One of Black's experiments was to set a pan of water on a steady fire and observe the temperature as a function of time. He found it steadily increased, reflecting the supply of heat from the fire, until the water began to boil, whereupon the temperature stayed the same for a long time. The steam coming off was at the same (boiling) temperature as the water.

So what was happening to the heat being supplied? Black correctly concluded that heat needed to be supplied to change water from its liquid state to its gaseous state, that is, to steam. In fact, a lot of heat had to be supplied: 540 calories per gram, as opposed to the mere 100 calories per gram needed to bring it from the freezing temperature to boiling.

He also discovered that it took 80 calories per gram to melt ice into water, with no rise in temperature. This heat is released when the water freezes back to ice, so it is somehow "hidden" in the water. He called it latent heat, meaning hidden heat.

THE GAS LAW

Coefficients of Expansion

Almost all materials expand on heating—the most famous exception being water, which contracts as it is warmed from 0 degrees Celsius to 4 degrees. This is actually a good thing, because as freezing weather sets in, the coldest water, which is about to freeze, is less dense than slightly warmer water, so rises to the top of a lake and the ice begins to form there. For almost all other liquids, solidification on cooling begins at the bottom of the container. So, since water behaves in this weird way, ice skating is possible! Also, as a matter of fact, life in lakes is possible—the ice layer that forms insulates the rest of the lake water from very cold air, so fish can make it through the winter.

LINEAR EXPANSION

The coefficient of linear expansion a of a given material, for example a bar of copper, at a given temperature is defined as the fractional increase in length that takes place on heating through one degree:

$$L \rightarrow L + \Delta L = (1+\alpha)\, L\, when T \rightarrow T + 1°C.$$

Of course, a might vary with temperature (it does for water, as we just mentioned) but in fact for most materials it stays close to constant over wide temperature ranges.

For copper, a = 17×10^{-6}

VOLUME EXPANSION

For liquids and gases, the natural measure of expansion is the coefficient of volume expansion, b.

$$V \rightarrow V + \Delta V = (1+\beta)V \text{ when } T \rightarrow T + 1°C.$$

Of course, on heating a bar of copper, clearly the volume as well as the length increases—the bar expands by an equal fraction in all directions (this could be experimentally verified, or you could just imagine a cube of copper, in which case all directions look the same).

The volume of a cube of copper of side *L* is *V* = *L*3. Suppose we heat it through one degree. Putting together the definitions of a, b above,

$$V \rightarrow (1+\beta)V, L \rightarrow (1+\alpha)L,\ L^3 \rightarrow (1+\alpha^3)\, L^3 or V \rightarrow (1+a)^3 V.$$

So $(1 + b) = (1 + a)^3$. But remember a is very, very small—so even though $(1 + a)^3 = 1 + 3a + a^2 + a^3$, the last two terms are completely negligible so to a fantastically good approximation:

$$b = 3a$$

The coefficient of volume expansion is just three times the coefficient of linear expansion.

GAS PRESSURE INCREASE WITH TEMPERATURE

In 1702, Amontons discovered a linear increase of P with T for air, and found P to increase about 33 per cent from the freezing point of water to the boiling point of water. That is to say, he discovered that if a container of air were to be sealed at 0 °C, at ordinary atmospheric pressure of 15 pounds per square inch, and then heated to 100 °C but kept at the same volume, the air would now exert a pressure of about 20 pounds per square inch on the sides of the container. Remarkably, Amontons discovered, if the gas were initially at a pressure of thirty pounds per square inch at 0 °C, on heating to 100 °C the pressure would go to about 40 pounds per square inch—so the percentage increase in pressure was the same for any initial pressure: on heating through 100 °C, the pressure would always increase by about 33 per cent.

Furthermore, the result turned out to be the same for different gases! Finding a Natural Temperature Scale.In class, we plotted air pressure as a function of temperature for a fixed volume of air, by making several measurements as the air was slowly heated (to give it a chance to all be at the same temperature at each stage). We found a straight line. On the graph, we extended the line backwards, to see how the pressure would presumably drop on cooling the air. We found the remarkable prediction that the pressure should drop to zero at a temperature of about –273 °C. In fact, if we'd done the cooling experiment, we would have found that air doesn't actually follow the line all the way down, but condenses to a liquid at around –200 °C.

However, helium gas stays a gas almost to –270 °C, and follows the line closely. The physics of gases, and the interpretation of this, much more fully in a couple of lectures. For now, the important point is that this suggests a much more natural temperature scale than the Celsius one: we should take -273 °C as the zero of temperature! For one thing, if we do that, the pressure/ temperature relationship for a gas becomes beautifully simple:

$$P \propto T.$$

This temperature scale, in which the degrees have the same size as in Celsius, is called the Kelvin or absolute scale. Temperatures are written 300 K. To get from Celsius to Kelvin, just add 273.15.

AN IDEAL GAS

Physicists at this point introduce the concept of an "Ideal Gas". This is like the idea of a frictionless surface: it doesn't exist in nature, but it is a very handy approximation to some real systems, and makes problems much easier to handle mathematically. The ideal gas is one for which $P\mu T$ for all temperatures, so helium is close to ideal over a very wide range, and air is close to ideal at ordinary atmospheric temperatures and above.

THE GAS LAW

We say earlier in the course that for a gas at constant temperature PV = constant (Boyle's Law). Now at constant volume,

$$P \mu T$$

We can put these together in one equation to find a relationship between pressure, volume and temperature:

$$PV = CT$$

where C is a constant. Notice, by the way, that we can immediately conclude that at fixed pressure, $V \mu T$, this is called Charles' Law. But what is C? Obviously, it depends on how much gas we have—double the amount of gas, keeping the pressure and temperature the same, and the volume will be doubled, so C will be doubled.

But notice that C will not depend on what gas we are talking about: if we have two separate one-liter containers, one filled with hydrogen, the other with oxygen, both at atmospheric pressure, and both at the same temperature, then C will be the same for both of them. One might conclude from this that C should be defined for one liter of gas at a specified temperature and pressure, such as 0 °C and 1 atmosphere, and that could be a consistent scheme. It might seem more natural, though, to specify a particular mass of gas, since then we wouldn't have to specify a particular temperature and pressure in the definition of C.

But that idea brings up a further problem: one gram of oxygen takes up a lot less room than one gram of hydrogen. Since we've just seen that choosing the same volume for the two gases gives the same constant C for the two gases, evidently taking the same mass of the two gases will give different C's.

AVOGADRO'S HYPOTHESIS

The resolution to this difficulty is based on a remarkable discovery the chemists made two hundred years or so ago: they found that one liter of nitrogen could react with exactly one liter of oxygen to produce exactly two liters of NO, nitrous oxide, all volume measurements being at the same temperature and pressure. Further, one liter of oxygen combined with two liters of hydrogen to produce two liters of steam.

These simple ratios of interacting gases could be understood if one imagined the atoms combining to form molecules, and made the further assumption, known as Avogadro's Hypothesis :

Equal volumes of gases at the same temperature and pressure contain the same number of molecules

One could then understand the simple volume results by assuming the gases were made of diatomic molecules, H_2, N_2, O_2 and the chemical reactions were just molecular recombinations given by the equations $N_2 + O_2 = 2NO$, $2H_2 + O_2 = 2H_2O$, etc. Of course, in 1811 Avogadro didn't have the slightest idea what this number of molecules was for, say, one liter, and nobody else did either, for another fifty years. So no-one knew what an atom or molecule weighed, but assuming that chemical reactions were atoms combining into molecules, or rearranging from one molecular pairing or grouping to another, they could figure out the relative weights of atoms, such as an oxygen atom had mass 16 times that of a hydrogen atom—even though they had no idea how big these masses were!

This observation led to defining the natural mass of a gas for setting the value of the constant C in the gas law to be a "mole" of gas: hydrogen was known to be H_2 molecules, so a mole of hydrogen was 2 grams, oxygen was O_2, so a mole of oxygen was 32 grams, and so on. With this definition, a mole of oxygen contains the same number of molecules as a mole of hydrogen: so at the same temperature and pressure, they will occupy the same volume. At 0 °C, and atmospheric pressure, the volume is 22.4 liters. So, for one mole of a gas (for example, two grams of hydrogen), we set the constant C equal to R, known as the universal gas constant, and equal to 8.3 J/(mol.K) and PV = RT.

For n moles of a gas, such as 2n grams of hydrogen, the law is:

$$PV = nRT$$

and this is the standard form of the Gas Law.

By the late 1700's, the experiments of Fahrenheit, Black and others had established a systematic, quantitative way of measuring temperatures, heat flows and heat capacities—but this didn't really throw any new light on just what was flowing.

This was a time when the study of electricity was all the rage, led in America by Benjamin Franklin, who had suggested in 1747 that electricity was one (invisible) fluid (it had previously been suggested that there were two fluids, corresponding to the two kinds of electrical charging observed).

CALORIMETRY

Calorimetry is the measurement of the quantity of heat exchanged. For example, if the energy from an exothermic chemical reaction is absorbed in a container of water, the change in temperature of the water provides a measure of the amount of heat added. Calorimeters are used to determine the energy content of foods by burning the foods in an oxygen atmosphere and measuring the energy yield in terms of the increase in temperature of the calorimeter. Calorimeters can also be used to measure the specific heat of a substance.

2

Combined Modes of Heat Transfer

INTRODUCTION

Experiments were conducted for ammonia-water falling film absorption in a plate heat exchanger with offset strip fins. The objectives of this paper were to analyse combined heat and mass transfer during the ammonia-water absorption process under different inlet subcooling modes, and to obtain heat transfer coefficients (Nusselt number).

This paper examined the effects of the inlet subcooling modes, the inlet concentration difference, liquid Reynolds number, and vapour Reynolds number on the heat transfer performance. Inlet liquid concentrations were set at 0, 5, 10, and 15 perc ent in mass of ammonia, while inlet vapour concentration ranged from 64.7 to 83.6 perc ent. Experiments were conducted in three ways according to the inlet subcooling conditions, *i.e.,* Case A (Tv>Tl), Case B (Tv~Tl), and CaseC (Tv<Tl). In Case A, there was a rectification process at the top of the test section by the inlet subcooling effect. Water desorption was confirmed in the experiments, which resulted in a lower absorption performance.

The heat transfer coefficient increased as the inlet subcooling increased in all cases. The effect of inlet subcooling on heat transfer performance was moresignificant in Case A than in Cases B and C. The inlet subcooling had more significant effect on the heat transfer performance than the inlet concentration difference. Nusselt number increased as liquid and vapour Reynolds numbers increased. The vapour velocity should be maximized to increase absorption performance in cocurrent ammonia-water absorption process. The parametric analysis provides fundamental understandings of the ammonia-water absorption process, and thus gives a guideline for heatexchanger compactness in ammonia-water absorption systems.

CONCEPT OF CONDUCTION

Heat can be conducted between two bodies which are in contact with each other; heat "flows" from one to the other.

- Materials which conduct heat well are called conductors of heat. Electrical conductors (such as metals) are good conductors of heat.
- Materials which do not conduct heat well are called insulators. Electrical insulators (for example, wood or glass) are usually good insulators of heat. Materials with low density, such as air or foamed plastic, are normally also good insulators unless they happen to be electrical conductors. To prevent heat from moving from one place to another, we usually place an insulator between.

Once a good insulator becomes hot, however, it stays that way for a long time, because it is difficult for the material to lose heat by conduction. Think of a hot ceramic pan and a hot metal pan: which cools faster?

Convection

This is a different kind of heat transfer than conduction. In conduction, heat itself is moving; in convection, hot portions of a fluid move through the body of the fluid. The hot fluid mixes with the cold fluid, and heat is transferred more quickly than by conduction.

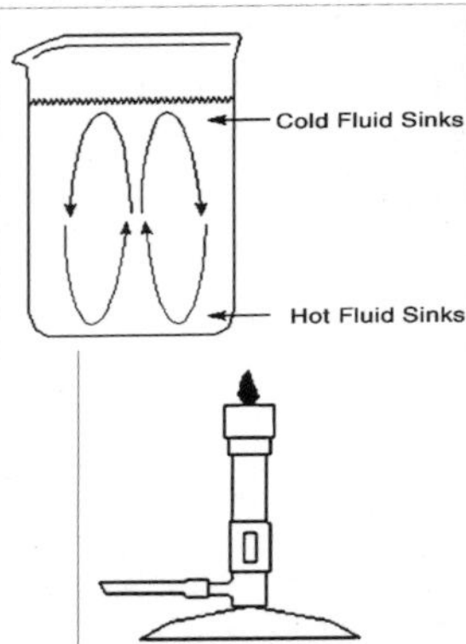

What we commonly call a "rolling boil" results from convection. Hot fluids rise through surrounding, cooler fluid because they are less dense; cooler fluids sink through warmer fluids because they are more dense. This causes circular motion of the fluid away from a source of heat. Convection in water drives ocean currents; convection in air drives weather patterns; and convection of molten rock inside the earth is thought to drive plate tectonics.

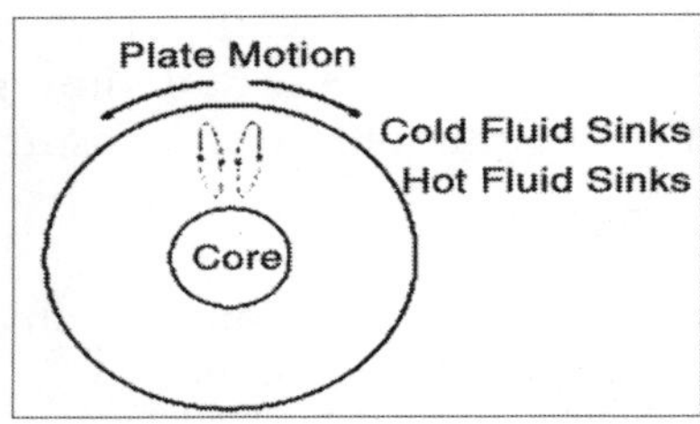

Convection leads to the counterintuitive fact that good insulators (like air) can transfer heat efficiently — as long as the air is allowed to move freely. Trapped air, as between panes of a double window, cannot transfer heat well because it cannot mix with air of a different temperature.

Radiation

Radiation is the simplest means of heat transfer. Heat radiation is carried not by moving atoms (as in conduction or convection) but by electromagnetic waves. Radiation is the only way that heat can move through a vacuum, and is the reason that even a closed thermos bottle (which has a vacuum between the inner and outer parts) will eventually come to the same temperature as its surroundings.

Heat transfer is most efficient by convection, then by conduction; radiation is the least efficient and slowest means of heat transfer. Low efficiency of heat transfer means that vacuums make excellent insulation.

COMPOSITE BODIES OF HEAT TRANSFER

Two bodies brought in thermal contact will change their temperature until they are at the same temperature. In the process of reaching thermal equilibrium,heat is transferred from one body to the other.

Suppose we have a system of interest at temperature TS surrounded by an environment with temperature TE. If TS > TE heat flows from the system to the environment. If TS < TE heat flows from the environment into the system. Heat, presented by the symbol Q and unit Joule, is chosen to be positive when heat flows into the system, and negative if heat flows out of the system. Heat flow is a results of a temperature difference between two bodies, and the flow of heat is zero if TS = TE.

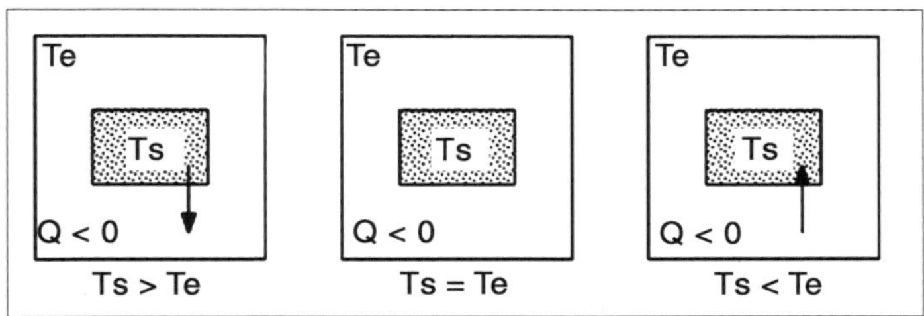

Fig. Heat Flow.

Heat is not the only way in which energy can be transferred between a system and its environment. Energy can also be transferred between a system and its environment by means of work (W). The unit of work is the Joule.

Another commonly used unit is the calorie. The calorie is defined as the amount of heat that would raise the temperature of 1 g of water from 14.5 ϒC to 15.5 ϒC. The Joule and the calorie are related as follows: 1 cal = 4.1860 J. When heat is added to an object, its temperature increases. The change in the temperature is proportional to the amount of heat added,

$$Q = C\left(T_f - T_i\right)$$

The constant C is called the heat capacity of the object. The heat capacity of an object depends on its mass and the type of material of which it is made. The heat capacity of an object is proportional to its mass, and the heat capacity per unit mass, c, is commonly used. In that case,

$$Q = cm\left(T_f - T_i\right)$$

where m is the mass of the object. The molar heat capacity is the heat capacity per mole of material. For most materials the molar heat capacity is 25 J/mol K.

In order to determine the heat capacity of a substance we not only need to know how much heat is added, but also the conditions under which the heat transfer took place. For gases, adding heat under constant pressure and under constant temperature will lead to very different values of the specific heat capacity.

When heat is added to a solid or a liquid, the temperature of the sample does not necessarily rise. During a phase change (melting, boiling) heat is added to the sample without an increase in temperature. The amount of heat transferred per mass unit during a phase change is called the heat of transformation (symbol L) for the process. The amount of heat needed/ released is given by,

$$Q = Lm$$

where m is the mass of the sample.

PROBLEM

What mass of steam of 100?C must be mixed with 150 g of ice at 0ΥC, in a thermally insulated container, to produce liquid water at 50ΥC ?

We start with calculating the heat required to transform 150 g of ice at 0ΥC to 150 g of liquid at 0ΥC. The heat of transformation of water is 333 kJ/ kg. The transformation of ice into water therefore requires a total heat given by,

$$Q = Lm = 50 \text{ kJ}.$$

The heat required to change the temperature of 150 g of water from 0ΥC to 50ΥC is given by,

$$Q = cm\,(T_f - T_i) = 31.5 \text{ kJ}.$$

The total heat that needs to be added to the system is therefore equal to 81.5 kJ. This heat must be supplied by the steam. Heat will be released when the steam is transformed into liquid, The heat of transformation for this process is 2260 kJ/kg. Suppose the mass of the steam is m. The total heat released in the conversion of steam into water is given by,

$$Q = Lm = 2260 \text{ mkJ}.$$

The heat released when the steam cools down from 100ϒC to 50ϒC is given by,

$$Q = cm\,(T_f - T_i) = 210 \text{ mkJ}.$$

The total heat released by the cooling of the steam is therefore equal to 2470 m kJ. The total heat required is 81.5 kJ, and we therefore conclude that the mass of the steam must be equal to 33 g.

Problem

A container of water has been outdoors in cold weather until a 5.0 cm-thick slab of ice has formed on its surface. The air above the ice is at –10ϒC. Calculate the rate of formation of ice (in centimeters per hour) on the bottom surface of the ice slab. Take the thermal conductivity of ice to be 0.0040 cal/ s. cm.ϒC and the density to be 0.92 g/cm3.

The water at the boundary between the water and the ice will be at a temperature of 0ϒC. The heat transported through 5 cm of ice is equal to

$$H = \frac{Q}{t} = kA\frac{T_H - T_C}{L} = 0.0080\,A\,cal/s$$

This heat is released when water is transformed into ice. The heat of transformation of this process is 79.5 cal/g. Suppose a mass m of water is transformed each second into ice. This produces a total heat equal to

H = 79.5 m cal/s

This must be equal to the heat flow through the ice:

79.5 m = 0.0080 A

A mass m of ice (covering an area A) will have a thickness d, where d is given by

$$d = \frac{m}{0.92A}$$

Combining the last two expression we obtain for the rate of ice formation:

$$d = \frac{m}{0.92A} = 1 \times 10^{-4}\,cm/s = 0.36\,cm/h.$$

METHODS OF HEAT TRANSFER

Temperature is a measure of the average amount of kinetic energy possessed by the particles in a sample of matter. The more the particles vibrate, translate and rotate, the greater the temperature of the object. You have hopefully adopted an understanding of heat as a flow of energy from a higher temperature object to a lower temperature object. It is the temperature difference between the two neighbouring objects that causes this heat transfer.

The heat transfer continues until the two objects have reached thermal equilibrium and are at the same temperature.

The discussion of heat transfer has been structured around some everyday examples such as the cooling of a hot mug of coffee and the warming of a cold can of pop. Finally, we have explored a thought experiment in which a metal can containing hot water is placed within a Styrofoam cup containing cold water. Heat is transferred from the hot water to the cold water until both samples have the same temperature.

A Particulate View of Transmission

Heat is transferred from the hot water to the cold water until both samples have the same temperature. In this instance, the transfer of heat from the hot water through the metal can to the cold water is sometimes referred to as conduction. Conductive heat flow involves the transfer of heat from one location to another in the absence of any material flow. There is nothing physical or material moving from the hot water to the cold water. Only energy is transferred from the hot water to the cold water. Other than the loss of energy, there is nothing else escaping from the hot water. And other than the gain of energy, there is nothing else entering the cold water. How does this happen? What is the mechanism that makes conductive heat flow possible?

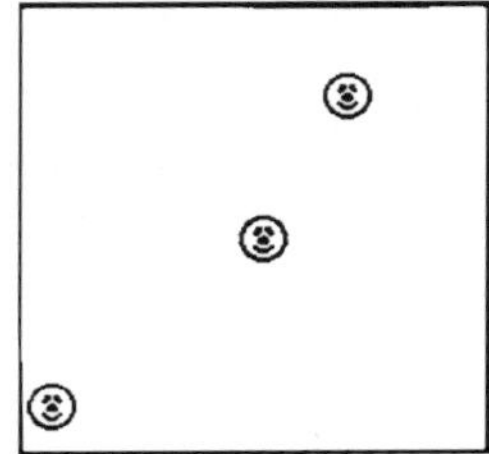

We have to think about matter as consisting of tiny particles atoms, molecules and ions. These particles are in constant motion; this gives them kinetic energy. These particles move throughout the space of a container, colliding with each other and with the walls of their container. This is known as translational kinetic energy and is the main form of kinetic energy for gases and liquids. But these particles can also vibrate about a fixed position. This gives the particles vibrational kinetic energy and is the main form of kinetic energy for solids. To put it more simply, matter consists of little wigglers and little bangers. The wigglers are those particles vibrating about a fixed position. They possess vibrational kinetic energy. The bangers are those particles that move through the container with translational kinetic energy and collide with the container walls.

The container walls represent the perimeters of a sample of matter. Just as the *perimeter of your property* (as in real estate property) is the furthest extension of the property, so the perimeter of an object is the furthest extension of the particles within a sample of matter. At the perimeter, the *little bangers* are colliding with particles of another substance-the particles of the container or even the surrounding air. Even the wigglers that are fixed in a position along the perimeter are doing some banging. Being at the perimeter, their wiggling results in collisions with the particles that are next to them; these are the particles of the container or of the surrounding air.

At this perimeter or boundary, the collisions of the little bangers and wigglers are elastic collisions in which the total amount of kinetic energy of all colliding particles is conserved. The net effect of these elastic collisions is that there is a transfer of kinetic energy across the boundary to the particles on the opposite side.

The more energetic particles will lose a little kinetic energy and the less energetic particles will gain a little kinetic energy. Temperature is a measure of the average amount of kinetic energy possessed by the particles in a sample of matter. So on average, there are more particles in the higher temperature object with greater kinetic energy than there are in the lower temperature object.

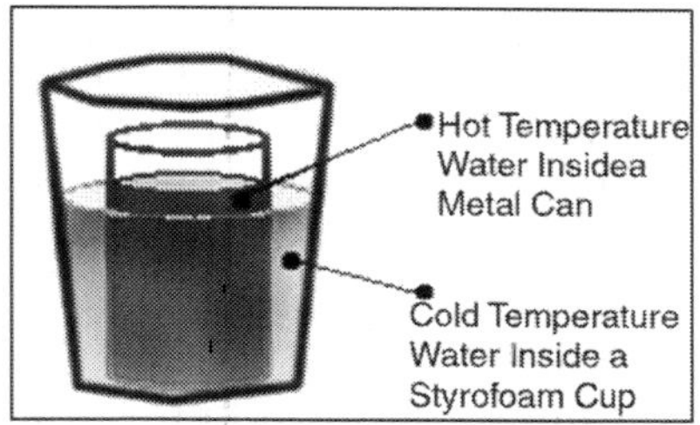

So when we average all the collisions together and apply the principles associated with elastic collisions to the particles within a sample of matter, it is logical to conclude that the higher temperature object will lose some kinetic energy and the lower temperature object will gain some kinetic energy. The collisions of our little bangers and wigglers will continue to transfer energy until the temperatures of the two objects are identical. When this state of thermal equilibrium has been reached, the average kinetic energy of both objects' particles is equal.

At thermal equilibrium, there are an equal number of collisions resulting in an energy gain as there are collisions resulting in an energy loss. On average, there is no net energy transfer resulting from the collisions of particles at the perimeter.

At the macroscopic level, heat is the transfer of energy from the high temperature object to the low temperature object. At the particulate level, heat flow can be explained in terms of the net effect of the collisions of a *whole bunch of little bangers*. Warming and cooling is the macroscopic result of this particle-level phenomenon. Now let's apply this particle view to the scenario of the metal can with the hot water positioned inside of a Styrofoam cup containing cold water.

On average, the particles with the greatest kinetic energy are the particles of the hot water. Being a fluid, those particles move about with translational kinetic energy and *bang upon* the particles of the metal can. As the hot water particles bang upon the particles of the metal can, they transfer energy to the metal can. This warms the metal can up. Most metals are good thermal conductors so they warm up quite quickly throughout the bulk of the can.

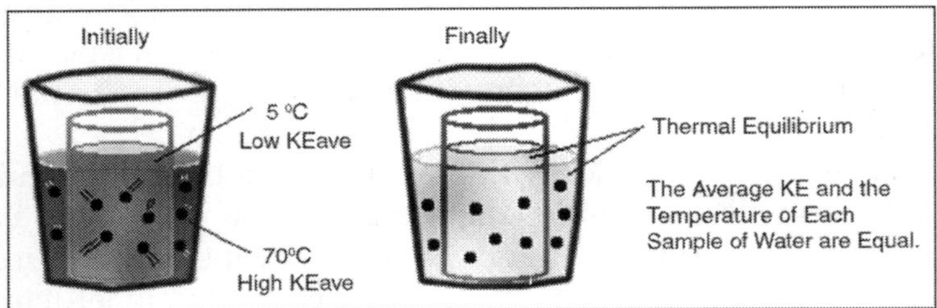

The can assumes nearly the same temperature as the hot water. Being a solid, the metal can consists of *little wigglers*. The wigglers at the outer perimeter of the metal can *bang upon* particles in the cold water. The collisions between the particles of the metal can and the particles of the cold water result in the transfer of energy to the cold water. This slowly warms the cold water up. The interaction between the particles of the hot water, the metal can and the cold water results in a transfer of energy outward from the hot water to the cold water.

The average kinetic energy of the hot water particles gradually decreases; the average kinetic energy of the cold-water particles gradually increases; and eventually, thermal equilibrium would be reached at the point that the particles of the hot water and the cold water have the same average kinetic energy. At the macroscopic level, one would observe a decrease in temperature of the hot water and an increase in temperature of the cold water.

The mechanism in which heat is transferred from one object to another object through particle collisions is known as conduction. In conduction, there is no net transfer of physical *stuff* between the objects. Nothing material moves across the boundary. The changes in temperature are wholly explained as the result of the gains and losses of kinetic energy during collisions.

Transmission Through the Bulk of an Object

The heat transfers from one object to another through conduction. But how does it transfer through the bulk of an object? For instance, suppose we pull a ceramic coffee mug out of the cupboard and place it on the countertop. The mug is at room temperature-maybe at 26°C. Then suppose we fill the ceramic coffee mug with hot coffee at a temperature of 80°C. The mug quickly warms up. Energy first flows into the particles at the boundary between the hot coffee and the ceramic mug. But then it flows through the bulk of the ceramic to all parts of the ceramic mug. How does heat conduction occur in the ceramic itself? The mechanism of heat transfer through the bulk of the ceramic mug is described in a similar manner as it before. The ceramic mug consists of a collection of orderly arranged wigglers. These are particles that wiggle about a fixed position.

As the ceramic particles at the boundary between the hot coffee and the mug warm up, they attain a kinetic energy that is much higher than their neighbours. As they wiggle more vigorously, they *bang into* their neighbours and increase their vibrational kinetic energy. These particles in turn begin to wiggle more vigorously and their collisions with their neighbours increase their vibrational kinetic energy.

The process of energy transfer by means of the little *bangers* continues from the particles at the inside of the mug (in contact with the coffee particles) to the outside of the mug (in contact with the surrounding air). Soon the entire coffee mug is warm and your hand feels it.

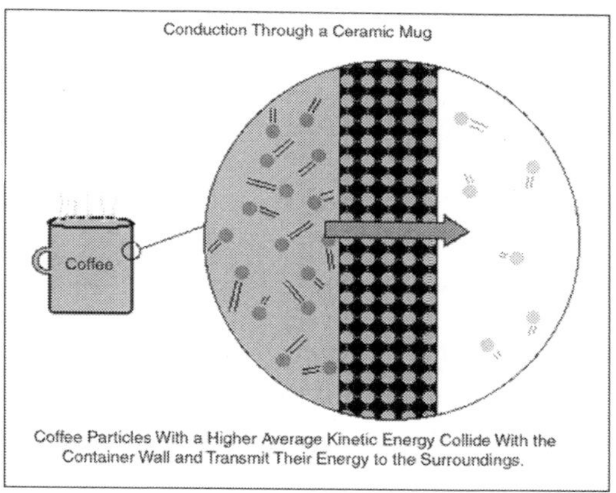

Coffee Particles With a Higher Average Kinetic Energy Collide With the Container Wall and Transmit Their Energy to the Surroundings.

This device of conduction by particle-to-particle interaction is very common in ceramic materials such as a coffee mug. Does it work the same in metal objects? For instance, you likely have noticed the high temperatures attained by the metal handle of a skillet when placed upon a stovetop. The burners on the stove transfer heat to the metal skillet. If the handle of the skillet is metallic, it too attains a high temperature, certainly high enough to cause a bad burn.

The transfer of heat from the skillet to the skillet handle occurs by conduction. But in metals, the conduction mechanism is slightly more complicated. In a manner similar to electrical conductivity, thermal conductivity in metals occurs by the movement of *free electrons*. Outer shell electrons of metal atoms are shared among atoms and are free to move throughout the bulk of the metal. These electrons carry the energy from the skillet to the skillet handle.

The details of this mechanism of thermal conduction in metals are considerably more complex. The main point to grasp is that heat transfer through metals occurs without any movement of atoms from the skillet to the skillet handle. This qualifies the heat transfer as being categorized as thermal conduction.

HEAT MOVE BY CONVECTION

The model of heat transfer through the ceramic coffee mug and the metal skillet involved conduction. The ceramic of the coffee mug and the metal of the skillet are both solids. Heat transfer through solids occurs by conduction. This is primarily due to the fact that solids have orderly arrangements of particles that are fixed in place.

Liquids and gases are not very good conductors of heat. In fact, they are considered good thermal insulators. Heat typically does not flow through liquids and gases by means of conduction. Liquids and gases are fluids; their particles are not fixed in place; they move about the bulk of the sample of matter. The model used for explaining heat transfer through the bulk of liquids and gases involves convection.

Convection is the process of heat transfer from one location to the next by the movement of fluids. The moving fluid carries energy with it. The fluid flows from a high temperature location to a low temperature location.

To understand convection in fluids, let's consider the heat transfer through the water that is being heated in a pot on a stove. Of course the source of the heat is the stove burner. The metal pot that holds the water is heated by the stove burner.

As the metal becomes hot, it begins to conduct heat to the water. The water at the boundary with the metal pan becomes hot. Fluids expand when heated and become less dense. So as the water at the bottom of the pot becomes hot, its density decreases. Differences in water density between the bottom of the pot and the top of the pot results in the gradual formation of *circulation currents*. Hot water begins to rise to the top of the pot displacing the colder water that was originally there.

And the colder water that was present at the top of the pot moves towards the bottom of the pot where it is heated and begins to rise. These circulation currents slowly develop over time, providing the pathway for heated water to transfer energy from the bottom of the pot to the surface.

Convection also explains how an electric heater placed on the floor of a cold room warms up the air in the room. Air present near the coils of the heater warm up. As the air warms up, it expands, becomes less dense and begins to rise. As the hot air rises, it pushes some of the cold air near the top of the room out of the way. The cold air moves towards the bottom of the room to replace the hot air that has risen. As the colder air approaches the heater at the bottom of the room, it becomes warmed by the heater and begins to rise. Once more, convection currents are slowly formed. Air travels along these pathways, carrying energy with it from the heater throughout the room. Convection is the main method of heat transfer in fluids such as water and air. It is often said that *heat rises* in these situations. The more appropriate explanation is to say that *heated fluid rises*. For instance, as the heated air rises from the heater on a floor, it carries more energetic particles with it.

As the more energetic particles of the heated air mix with the cooler air near the ceiling, the average kinetic energy of the air near the top of the room increases. This increase in the average kinetic energy corresponds to an increase in temperature. The net result of the rising hot fluid is the transfer of heat from one location to another location.

The convection method of heat transfer always involves the transfer of heat by the movement of matter. This is not to be confused with the caloric theory. In caloric theory, heat was the fluid and the fluid that moved was the heat. Our model of convection considers heat to be energy transfer that is simply the result of the movement of more energetic particles.

The two examples of convection — heating water in a pot and heating air in a room-are examples of natural convection. The *driving force* of the circulation of fluid is natural-differences in density between two locations as the result of fluid being heated at some source. Natural convection is common in nature.

The earth's oceans and atmosphere are heated by natural convection. In contrast to natural convection, forced convection involves fluid being forced from one location to another by fans, pumps and other devices. Many home heating systems involve force air heating. Air is heated at a furnace and blown by fans through ductwork and released into rooms at vent locations. This is an example of forced convection. The movement of the fluid from the hot location (near the furnace) to the cool location (the rooms throughout the house) is driven or forced by a fan.

Some ovens are forced convection ovens; they have fans that blow heated air from a heat source into the oven. Some fireplaces enhance the heating ability of the fire by blowing heated air from the fireplace unit into the adjacent room. This is another example of forced convection.

HEAT TRANSFER BY RADIATION

A final method of heat transfer involves radiation. Radiation is the transfer of heat by means of electromagnetic waves. To *radiate* means to send out or spread from a central location. Whether it is light, sound, waves, rays, flower petals, wheel spokes or pain, if something *radiates* then it protrudes or spreads outward from an origin.

The transfer of heat by radiation involves the carrying of energy from an origin to the space surrounding it. The energy is carried by electromagnetic waves and does not involve the movement or the interaction of matter. Thermal radiation can occur through matter or through a region of space that is void of matter (*i.e.*, a vacuum). In fact, the heat received on Earth from the sun is the result of electromagnetic waves travelling through the *void of space* between the Earth and the sun.

All objects radiate energy in the form of electromagnetic waves. The rate at which this energy is released is proportional to the Kelvin temperature (T) raised to the fourth power.

Radiation rate = $k \bullet T^4$

The hotter the object, the more it radiates. The sun obviously radiates off more energy than a hot mug of coffee. The temperature also affects the wavelength and frequency of the radiated waves. Objects at typical room temperatures radiate energy as infrared waves. Being invisible to the human eye, we do not see this form of radiation. An infrared camera is capable of detecting such radiation.

Perhaps you have seen thermal photographs or videos of the radiation surrounding a person or animal or a hot mug of coffee or the Earth. The energy radiated from an object is usually a collection or range of wavelengths. This is usually referred to as an *emission spectrum*. As the temperature of an object increases, the wavelengths within the spectra of the emitted radiation also decrease. Hotter objects tend to emit shorter wavelength, higher frequency radiation. The coils of an electric toaster are considerably hotter than room temperature and emit electromagnetic radiation in the visible spectrum. Fortunately, this provides a convenient warning to its users that the coils are hot.

The tungsten filament of an incandescent light bulb emits electromagnetic radiation in the visible (and beyond) range. This radiation not only allows us to see, it also warms the glass bulb that contains the filament. Put your hand near the bulb (without touching it) and you will feel the radiation from the bulb as well.

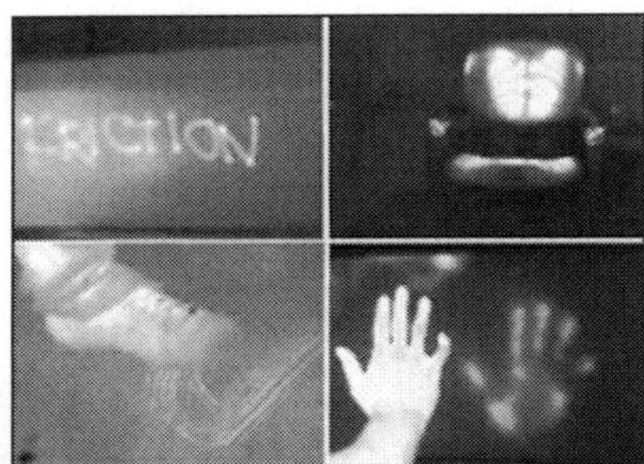

Thermal radiation is a form of heat transfer because the electromagnetic radiation emitted from the source carries energy away from the source to surrounding (or distant) objects. This energy is absorbed by those objects, causing the average kinetic energy of their particles to increase and causing the temperatures to rise. In this sense, energy is transferred from one location to another by means of electromagnetic radiation. The image at the right was taken by a thermal imaging camera.

The camera detects the radiation emitted by objects and represents it by means of a colour photograph. The *hotter* colours represent areas of objects that are emitting thermal radiation at a more intense rate.

MODES OF HEAT TRANSFER

CONDUCTION

- Energy transfer at a molecular level
- No physical movement of the material
- Heating/Cooling of the solid material

The rate of heat flux (rate of heat transfer per unit area) in a solid object is proportional to the temperature gradient, this can be stated mathematically as,

$$\frac{q_x}{A} \, \alpha \, \frac{dT}{dx}$$

We may remove the proportionality by using a constant 'k', to obtain, Fourier's Law

$$q_x = -kA\frac{dT}{dx}$$

where,

qx = rate of heat transfer in the x direction by conduction, W
k = thermal conductivity, W/mC
A = area (normal to x-direction) through which heat flows, m^2
T = temperature, C
x = length, variable, m
Sign Convention:

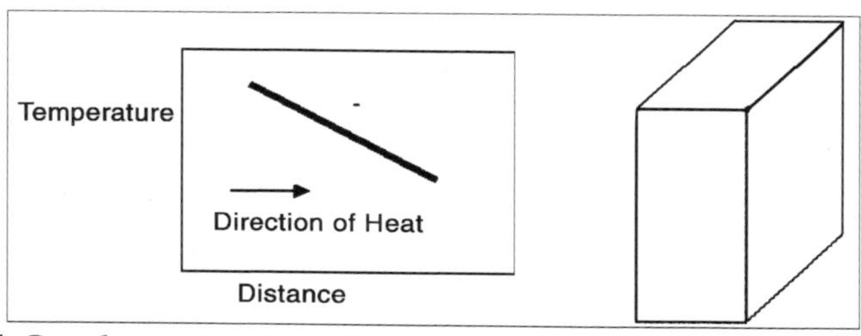

Thermal Conductivity, k unit: W/mC
Metals: k = 50–400 W/mC
Water: k = 0.597 W/mC
Air: k = 0.0251 W/mC
Insulating materials: k = 0.035 – 0.173 W/mC
For foods
k = 0.25 mc + 0.155 mp + 0.16 mf + 0.135 ma + 0.58 mm

Where m is mass fraction and subscripts c: carbohydrate, p: protein, f: fat, a: ash, m: moisture.

CONVECTION

Fluid flow over a solid body—heat transfer between a solid and a fluid.

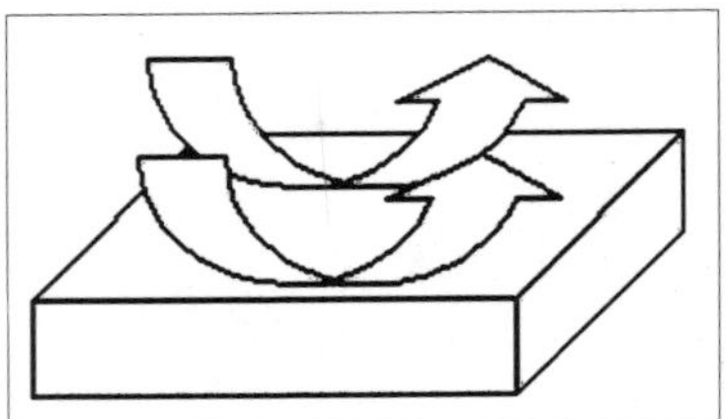

Newton's Law of Cooling:

$q = h\ A\ (T_p - T_\alpha)$

where: h is convective heat transfer coefficient (W/m^{2}C), A is area (m^2), T_p is plate surface temperature (°C), T_α is surrounding fluid temperature (°C).

- *Forced Convection*: Artificially induced fluid flow
- *Free (Natural) Convection*: Caused due to density differences

Fluid condition	h (W/m^{2}C)
Air, free convection	5-25
Air, forced convection	10-200
Water, free convection	20-100
Water, forced convection	50-10,000
Boiling water	3,000-100,000
Condensing water vapour	5,000-100,000

RADIATION

Heat transfer between two surfaces by emission and later absorption of electromagnetic radiation

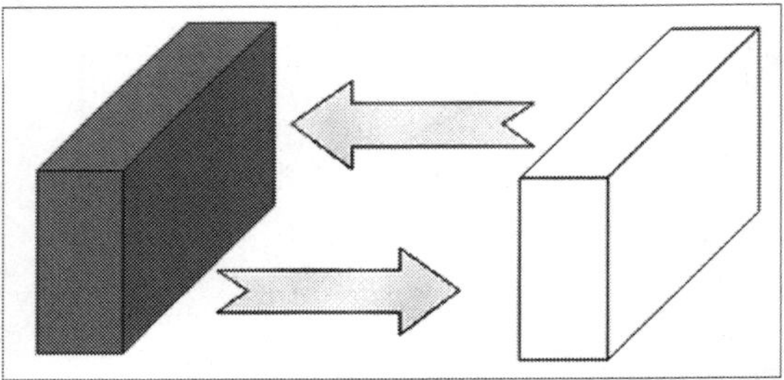

- Requires no physical medium
- Stefen-Boltzmann Equation:

$$q = A\ \sigma\ \varepsilon\ (T_2^4 - T_1^4)$$

where,

σ = Stefen-Boltzmann's constant, 5.669×10^{-8} W/m^{2}K^4

ε = emissivity, (varies from 0 to 1) dimensionless

A = area, m^2

T_1 = temperature of surface 1, Absolute

T_2 = temperature of surface 2, Absolute

FOURIER LAW OF HEAT CONDUCTION

Fourier's law is an empirical law based on observation. It states that the rate of heat flow, dQ/dt, through a homogeneous solid is directly proportional to the area, A, of the section at right angles to the direction of heat flow, and to the temperature difference along the path of heat flow, dT/dx *i.e.,*

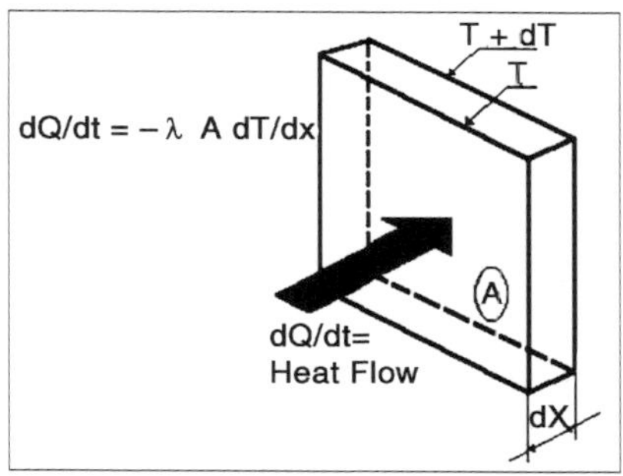

Up to this point little has been said about the physical laws determing the flow of heat $\vec{q}$ or the flux of salt $\vec{F_S}$ or the viscous stresses τ_{ij}. The laws governing these quantities represent additional hypotheses, the justification of which lies in the satisfactory comparison of deductions from these laws with experiment. Such comparisons between theory and observation are limited, particularly in respect to the more complex situations which may arise where several different fluxes exist simultaneously, and it is conceivable and even probable that the classical ``laws'' of heat conduction, diffusion, and momentum transfer must be altered to account for phenomena which cannot be otherwise explained.

In short, the state of our knowledge of molecular transfer of properties is far from complete. Nevertheless, there is one guide in the selection of the hypothesis governing the non-advective flux: it must be consistent with the second law of thermodynamics as set forth in the inequality. In order to understand the nature of the restrictions imposed by this law it is necessary to consider some examples. We will start with the simplest case.

Fourier showed that the flow of heat in a solid could be explained if,

$$\vec{q} = -C\Delta T$$

where C, the coefficient of thermal conductivity, is a property of the particular substance. There is evidence that this law holds in a pure fluid under certain conditions. The coefficient C may vary with the temperature T, and certainly varies from one substance to another, but the feature which it has in common with all substances according to empirical evidence is that it is always a positive constant.

This of course makes intuitive sense, at least if the molecular concept of temperature is invoked; the heat (kinetic energy at the microscopic scale in another sense) tends to flow from regions of high concentration of internal energy to regions of low internal energy, which is consistent with the statement above that the heat flow is directed in the direction of the gradient of the temperature $-\nabla T$.

This is consistent with the second law of thermodynamics. To show this we must make use of the first law in the form (VI-1) which, when compared with (V-2), indicates that the terms on the right hand side of (VI-1) must at least be zero. Now, since the above law for $\vec{q}$ is independent of velocity gradients or gradients of concentration such as would exist in a fluid mixture, it is evidence that the last term,

$$\frac{q}{T^2}.(-\nabla T)$$

must be at least zero regardless of what value the other terms may have, for these can be zero independently (and would be in a solid under the conditions of Fourier's experiments). Therefor, making use of Eq., it is required that

$$\frac{C}{T^2}(\nabla T)^2 \geq 0,$$

which in general requires that $C \geq 0$, and therefore is consistent with experiment.

NEWTION'S LAW OF COOLING FOR CONVECTION

At contact surfaces between a fluid and a solid wall, there is always a thin layer of fluid through which the heat is transferred by conduction. Whenever there is an appreciable movement of the fluid, conduction heat transfer in fluid may be neglected compared with convection heat transfer.

The heat transfer from the solid surface to the fluid can be described by Newton's law of cooling. It states that the heat transfer, dQ/dt, from a solid surface of area A, at a temperature Tw, to a fluid of temperature T, is:

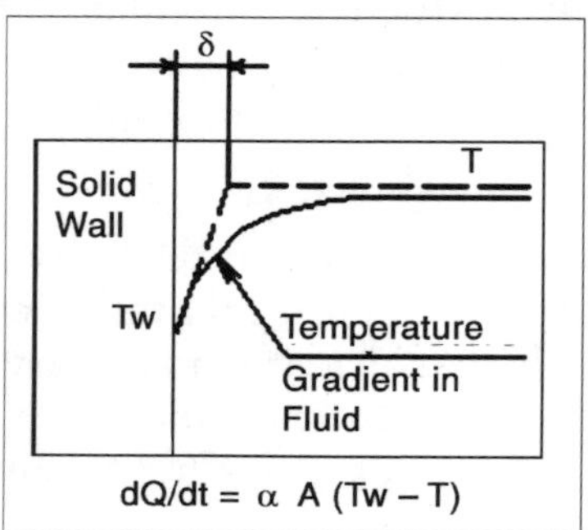

$$\dot{Q} = hA(T_w - T_\infty) = hA \cdot \Delta T$$

The rate of heat $\dot{Q}$ transfered to the surrounding fluid is proportional to the object's exposed area A, and the difference between the object temperature T_w and the fluid free-stream temperature T_∞. The constant of proportionality h is termed the convection heat-transfer coefficient. Other terms describing h include film coefficient and film conductance. Variables needed to parameterize free convection are:

DEFINITION OF SYMBOLS

Independent Parameters for the Fluid

Quantity	Symbol	Units
Bulk temperature		K
Kinematic viscosity	ν	m^2/s
Coef. of thermal expansion	β	1/K
Dynamic viscosity	μ	kg/m-s
Density	ρ	kg/m^3
Thermal diffusivity	α	m^2/s
Specific heat	c_p	J/kg-K
Thermal conductivity	k	W/m-K

Independent Parameters for the Object

Quantity	Symbol	Units
Surface reference length	L	m
Surface diameter (for pipes)	D	m
Surface (wall) temperature	T_w	K

Dependent Parameters

Quantity	Symbol	Units
Surface to fluid temp. difference		K
Heat transfer coefficient	h	W/m^2-K

CONCEPT OF THERMAL RADIATION

Thermal radiation is the energy radiated from hot surfaces as electromagnetic waves. It does not require medium for its propagation. Heat transfer by radiation occur between solid surfaces, although radiation from gases is also possible. Solids radiate over a wide range of wavelengths, while some gases emit and absorb radiation on certain wavelengths only.

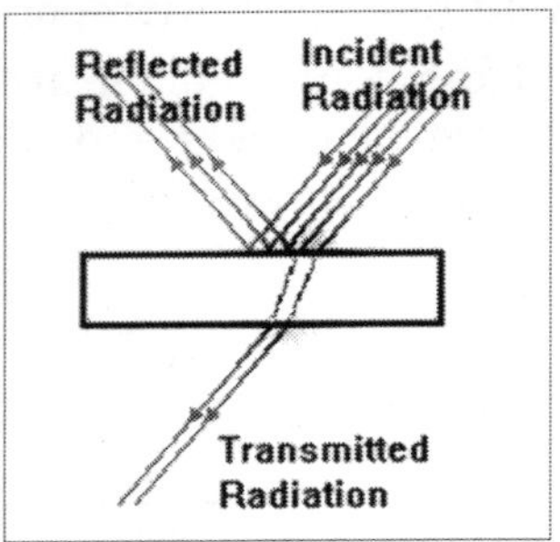

When thermal radiation strikes a body, it can be absorbed by the body, reflected from the body, or transmitted through the body. The fraction of the incident radiation which is absorbed by the body is called absorptivity (symbol α). Other fractions of incident radiation which are reflected and transmitted are called reflectivity (symbol ρ) and transmissivity (symbol τ), respectively. The sum of these fractions should be unity *i.e.*

$$\alpha + \rho + \tau = 1$$

All objects give off radiation, *i.e.,* electromagnetic waves, because of their temperature; we refer to this as thermal radiation. As well they absorb such radiation from their surroundings. If a body is hotter than its surroundings it emits more radiation than it absorbs, and tends to cool; if a body is cooler than its surroundings it absorbs more radiation than it emits, and tends to warm. Usually it will eventually come to thermal equilibrium with its surroundings: a condition in which its rates of absorbtion and emission of radiation are equal.

The spectrum of the thermal radiation from a hot solid body is continuous; that is, it emits electromagnetic radiation in a continuous range of wavelengths (or frequencies), over a rather large band of allowed wavelengths. It is usual for the amount of radiation at any given frequency to be different than at other frequencies, depending of course on the body, and the atoms that constitute it. We refer to the set of all these frequencies as its spectrum and are often interested not only in the range of frequencies but also the amount of radiation in some small frequency interval.

In particular it is interesting to look at the portion of the spectrum where "a majority" of the radiation is emitted by the body, *i.e.,* the "predominant" part of its radiation spectrum. It seems reasonable that this would have something to do with what the body is made of and also what is its temperature.

Most bodies with which we normally come into contact have the predominant portion of their radiation spectrum in the infrared, or far-infrared, portion of the spectrum, so that it is not visible to us. However, as the temperature of a body is increased more radiation is emitted, and also the dominant portion of its spectrum shifts so that its radiation begins to be visible to humans. We usually could describe this by saying that, as some object is heated more and more, the predominant colour of the hot object shifts from dull red through bright yellow-orange to bluish, "white heat."

As already mentioned, the radiation depend not only on the temperature but also the composition of the object. For example, at 2000 K a polished, flat tungsten surface emits radiation at a rate of 23.5 Watts/cm^2; for molybdenum, however, the corresponding rate is only 19.2 Watts/cm2. In each of these cases the rate increases somewhat if the surface is roughened, instead of being polished. Other differences appear if we measure the distribution in wavelengths of the emitted radiation. This dependence on materials makes the problem quite complicated. Hopefully it reminds us of some of the difficulties one has with describing gases of atoms.

Gases of different sorts of atoms have different properties; the more elementary parts of their behaviour could, however, be understood rather well by introducing, instead, the idea of an ideal gas, which was fairly easy to study, and could then be used as a beginning to understand more of the details of more complicated gases. This ideal gas was really an idealization, or abstraction, of the basic ideas surrounding the behaviour of gases, which was fairly straight-forward to study because of their simplicity. Likewise, in considering ``the radiation problem,'' it is useful to introduce the concept of "an ideal radiator." For such an object we assume that the spectrum of the emitted thermal radiation depends only on the temperature of the radiator and not on the material or the nature of the surface.

In the real world, a gas which behaves not very differently than an ideal gas can be obtained by considering a gas of monatomic atoms, such as helium, neon, etc., at temperatures that are not too high. Likewise, in the real world, one may obtain a very good approximation to an ideal radiator by considering the radiation emitted through a small hole of a hollow body, heated to some temperature. The radiation is emitted by the interior walls of the cavity, and becomes quite mixed before it manages to find the small hole to escape; as a result of this it depends very little on the actual material of the body, or its surface.

We therefore refer to this radiation as *cavity radiation*. We can also consider this situation in reverse, and consider the radiation that is absorbed--rather than emitted--by this hole. As it is simply a hole--with a dark hollow on the far side--it absorbs all radiation that strikes it, independent, again, of its material. Therefore, often such radiation is referred to, instead, as *black-body radiation*. We will now spend the rest of this material trying, therefore, to understand the details of the spectrum of black-body radiation, and its dependence upon temperature. Such studies were made near the beginning of the twentieth century, and shortly thereafter. The total radiated power per unit area of a cavity aperture, summed over all wavelengths, is called its *radiant intensity*, I(T). It depends on the temperature T, has units of Watt/m^2, and is given by a simple formula, given below, which involves σ, the Stefan-Boltzmann constant, which is a universal constant, independent of material, surface, or temperature.

We note, however, that ordinary hot objects always radiate less efficiently than do cavity radiators; therefore, we associate with them an additional, positive quantity, ε, referred to as the emissivity, which is always less than 1, and depends on the material, the surface, and (usually) the temperature as well. Although it is called the emissivity, it also is used to describe the absorption of radiation by that material as well:
for a blackbody, or cavity aperture:

$$I(T) = \sigma T^4,\ \sigma = 5.670 \times 10^{-8}\ \text{W/m}^2 - \text{K}^4,$$

for a real material:

$$I(T) = \varepsilon\sigma T^4.$$

ELECTRICAL CIRCUITS

ELECTRICAL ANALOGY

In the development of his law for electrical circuits, Georg Ohm performed experiments that modeled Fourier's law of heat conduction. Consequently, an analogy between heat and electrical conduction can be observed. Symbolically Ohm's law can be expressed as

$$I = \frac{\Delta V}{R_e}$$

where I is the current flowing through an element, ΔV is the voltage across the element, and Re is the electrical resistance across the element. Similarly Fourier's law can be written as,

$$q = \frac{\Delta T}{R_t}$$

where q is the rate of heat conduction, ΔT is the temperature difference between the surfaces of a slab, and Rt is the thermal resistance between the surfaces defined as,

$$R_t = \frac{L}{kA}$$

where k is the thermal conductivity and A is the effective area of the resistance. By using the convective heat transfer coefficient h, the convective thermal resistance from a surface to a fluid can also be calculated by the equation,

$$RConvective = \frac{1}{hA}$$

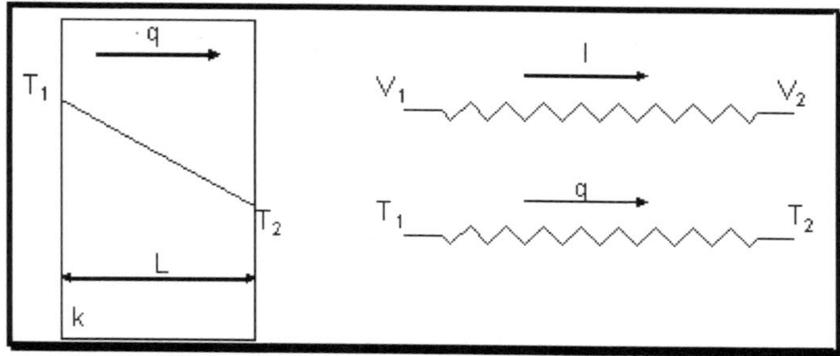

Since heat is understood to transfer through lattice vibrations between adjacent atoms, and electricity is conducted by free valence electrons of atoms, it may seem intuitive that heat and electrical conduction are analogous. Although this is most often true, some non metallic materials like diamond exhibit dissimilar heat and electrical conductivities.

USES OF THE THERMAL-ELECTRICAL ANALOGY

As found in the referenced literature, the analogy is useful in the analysis of several steady heat transfer problems from property measurement to modeling. In modeling, a complicated heat transfer analysis can be made much simpler by creating an "electric circuit" like model of the problem.

This can be seen in the illustrative examples below. In the first example, a simple comparison between single and triple paned windows is modeled. In the second example an entire house is modeled using a thermal network. Also evident from the illustrative examples is the fact that one should take additional care in determining if system components are acting in series or parallel.

LIMITS OF THE THERMAL-ELECTRICAL ANALOGY

Although generally extremely useful, the analogy is not without its own limitations. These include the non-linearities that occur between voltage and current at extremely high and low values.

The literature survey would also lead to special cases where the analogy doesn't quite fit the actual physics of a problem. Recall the constitutive models of heat transfer are based on the continuum concept. These special cases are at extremely large or small values of current and voltage as well as heat transfer problems at the nano-scale level. Despite this, modifications can be made to still find the relationship helpful in the analysis.

CONDUCTION HEAT TRANSFER

Conduction is the transfer of energy from a more energetic to the less energetic particles of substances due to interactions between the particles. In gas and liquids, heat conduction takes place through random molecular motions (diffusions), in solid heat conduction is through lattice waves induced by atomic motions.

In nonconductor, the energy transfer is exclusively through such lattice waves, in conductors, the motion of free electrons also contribute to heat conduction. Fourier Law is the rate equation based on experimental evidences. This law states that heat flux at the surface normal is proportional to the temperature gradient.

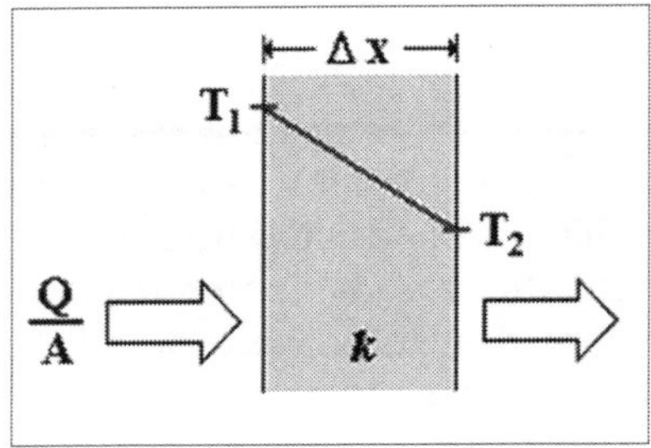

Fig. 1D heat transfer

For 1D heat conduction in x-direction:

$$q'' = -k\, dT/dx$$

Where k is thermal conductivity (W/m.K), T is temperature (K), q" is the heat flux in x direction.

General form of Fourier's Law is:

$$q'' = -k\nabla\ T(x,y,z,t) = -k(i\partial T/\partial x + j\partial T/\partial y + k\partial T/\partial z)$$

Thermal conductivity k is a material property; it may vary with temperature and pressure. In laser machining which usually involves solids, the dependence on temperature should be considered. This makes k a function of temperature k(T), such function relations are found through interpolation of experimental results at various temperatures. Using energy balance relations, ie., increase of stored energy in a volume of material = energy in + energy generation in this volume - energy out, we can deduce the general differential form of three dimensional heat conduction equation. In short:

$$Ein + Eg - Eout = \Delta\ Est$$

General equation of Heat Conduction:

$$\partial (k\partial T/\partial x)/\partial x + \partial (k\partial T/\partial y)/\partial y + \partial (k\partial T/\partial z)/\partial z + dq/dt = \rho\ cp\ (\partial T/\partial t)$$

T= T(x,y,z,t). Steady state:

$$(\partial T/\partial t) = 0$$

If material is of constant property, we can pull k out and we get:

$\partial^2 T/\partial x^2 + \partial^2 T/\partial y^2 + \partial^2 T/\partial z^2 + (dq/dt)/k = (\rho\, cp/k)(\partial T/\partial t) = 1/\alpha\,(\partial T/\partial t)$

$\alpha = k/(\rho\, cp)$ is the thermal diffusivity.

In cylindrical coordinates we have:

Fourier Law: $q'' = -k\,\nabla T(r,\phi,z,t) = -k(i\partial T/\partial r + j(1/r)\partial T/\partial \partial + k\partial T/\partial z)$

General equation of Heat Conduction:

$(1/r)\,\partial(k\partial T/\partial r)/\partial r + (1/r^2)\partial(k\partial T/\partial \phi)/\partial \phi + \partial(k\partial T/\partial z)/\partial z + dq/dt = \rho\, cp\,(\partial T/\partial t)$

Where r is the radial direction, $\phi\partial$ is the circumferential direction, z is the axial direction. When we assume axis symetric heat transfer, which happens frequently in laser drilling modeling, we mean that the variation of temperature in the $\phi\partial$ direction is zero. Then the above equation becomes:

$(1/r)\partial(k\partial T/\partial r)/\partial r + \partial(k\partial T/\partial z)/\partial z + dq/dt = \rho\, c_p\,(\partial T/\partial t)$

Similarly in spherical coordinates:

$q'' = -k\,\nabla T(r,\theta,\phi,t) = -k(i\partial T/\partial r + j(1/r)\partial T/\partial \theta + k(1/r \sin\phi)\partial T/\partial \phi)$

$(1/r^2)\,\partial(kr^2\partial T/\partial r)/\partial r + (1/r^2\sin^{2\theta})\,\partial(k\partial T/\partial \phi)/\partial \phi + (1/r^2 \sin\theta)\,\partial(k \sin\theta\, \partial T/\partial \theta)/\partial \theta + dq/dt = \rho\, cp\,(\partial T/\partial t)$

These equations are complex, but they still state the conservation of energy.

CONVECTION HEAT TRANSFER

Convection usually refers to the energy transfer between a solid surface and an adjacent moving gas or liquid. Convection heat transfer is a combination of diffusion or molecular motion within the fluid and the bulk or macroscopic motion of the fluid. The rate of energy transfer from a system to the fluid is quantified by Newton's law of cooling:

$$q'' = h(T_{fluid} - T_s)$$

where q" is the convective heat flux (W/m^2), h is the convective heat transfer coefficient (W/m^2K), Ts is the surface temperature (K) and

Tfluid is fluid temperature away from the surface (K). The heat transfer coefficient, h, is an empirical parameter that encompasses the effects of the fluid flow near the surface, the fluid properties, and the surface geometry. The decision of h is the most important task in convection heat transfer analysis. Many empirical formulations are put forward, readers should refer to textbook of Heat Transfer for detailed information. Convective heat transfer is coupled with the fluid dynamics of the system. The following figure show the boundary layer development of a uniform flow along a flat plate.

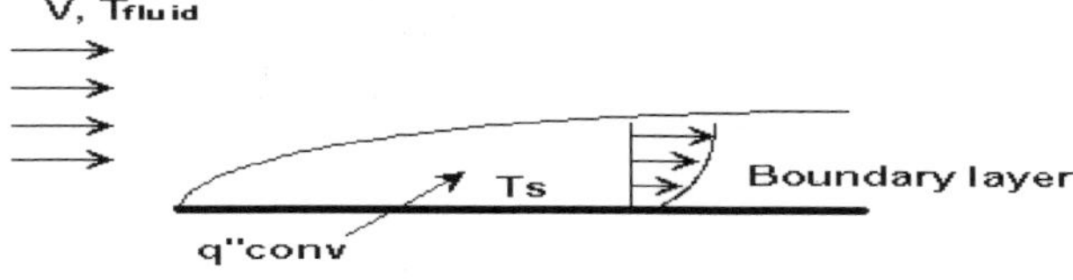

Fig. Illustration of Boundary Layer and Convection Heat Transfer

RADIATION HEAT TRANSFER

Unlike conduction and convection, radiation does not require the presence of a medium to propagate. Actually, radiation transfers heat energy most efficiently in a vacuum. Radiation energy is transported by electromagnetic waves. Thermal radiation is energy emitted by matter at a finite temperature as a result of changes in the electron configurations of the atoms or molecules. Generally, radiation heat transfer analysis is focused on solid surfaces but emission may also occur from liquids and gases. Stefan-Boltzmann Law states that emissive power E of a surface is proportional to T4 :

$$E = \varepsilon \sigma T_s^4,$$

Where ε is surface emissivity; $\sigma = 5.67 * 10^{-8} W/m^2.K^4$ is the Boltzmann constant. The energy absorbed by the surface is:

$$G_{abs} = \alpha G = \alpha \sigma T_{sur}^4,$$

where ε is absorptive; G is the irradiation from the surroundings.

So the net heat transfer of radiation energy from a finite surface to the infinite surrouding is:

$$q''_{rad} = E - G_{abs} = \varepsilon \sigma T_s^4 - \alpha \sigma T_{sur}^4$$

Where Ts and Tsur are the surface and surrounding temperatures.

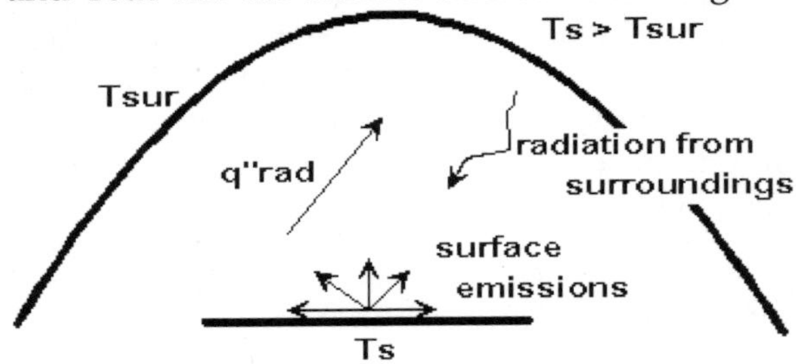

Fig. Radiation Energy Exchange between a Surface and the Surounding

The real analysis of radiation energy exchange is far more complex than what we see here. Radiation relates to direction, area, wavelength, surface condition, etc. We also see that radiation increases with 4th power of temperature, thus at high temperature the radiation heat transfer should be considered.

EULERIAN AND LAGRANGIAN DESCRIPTIONS

There are two basic coordinate systems which may be used in mechanics. In Eulerian framework the independent variables are the spatial coordinates x,y,z and time t. Attention is focused on the fluid that passes through a control volume which is fixed in space, the fluid inside the control volume at any instant in time will consist of different fluid particles from that which was there at some previous instance in time. In Lagrangian approach, attention is focused on a particular mass of fluid as it flows, in this frame, x,y,z,t are no longer independent variables, the independent variables are x_0, y_0, z_0 at time t_0 and time t. All positions of the mass we are considering can be calculated if the velocity field is known. Let F be any field variable such as density or temperature, F=F(x,y,z,t) in Eulerian description or F= F(x_0,y_0,z_0,t) in Lagrangian description. Then, the time derivative of F is DF/Dt, and we have:

$$\frac{DF}{Dt} = \frac{\partial F}{\partial t} + u\frac{\partial F}{\partial x} + v\frac{\partial F}{\partial y} + w\frac{\partial F}{\partial z},$$

where u, v, w are velocities in the x, y, z directions.

CONSERVATION OF MASS

Consider a specific mass of fluid whose volume V is arbitrarily chosen. This fluid mass is followed as it flows, its shape and size can change with time, but its mass will remain unchanged unless some mass generation/ extinction mechanism exists (nuclear reaction for example). So the conservation of mass states that the Lagrangian derivative of the mass of the fluid inside the volume is zero. In mathematical form:

$$\frac{D}{Dt}\int_v p dV = 0$$

(Integration form)

The PDE form of conservation of mass is:

$$\frac{Dp}{dt} + \frac{\partial}{\partial x_k}(puK) = 0$$

where xk and uk are coordinates and velocity in the k = 1,2,3 directions.

If the fluid is incompressible, then we have:

$$\frac{Dp}{dt} = 0$$

Combine this relation with the general law of mass conservation, we have the well known Continuity Equation for incompressible flows:

$$\frac{\partial u}{\partial x} + \frac{\partial v}{\partial y} + \frac{\partial w}{\partial z} = \nabla \vec{V} = 0$$

CONSERVATION OF MOMENTUM

The principle of conservation of momentum is an application of Newton's second law of motion to an element of the fluid. Considering a given mass of fluid in a Lagrangian frame, this law states that the rate at which the momentum of the fluid mass is changing is equal to the net external force acting on the mass. The external forces may be classified as body forces (such as gravitational or electromagnetic forces) $\vec{f}$ and surface forces (such as pressure forces or viscous forces) $\vec{p}$. Thus we have:

$$\frac{D}{Dt}\int_V p\vec{V}dV = \int_s \vec{P}ds + \int_V p\vec{f}dV$$

After many manipulations, using continuity relation and Renolds' Transport Theorem, and relate the surface force to the stress tensor σ_{ij} and surface normal ni, (i = 1,2,3 , j = 1,2,3). We got the Momentum equations:

$$p\frac{\partial u_j}{\partial t} + pu_k\frac{\partial u_j}{\partial x_k} = \frac{\partial \sigma_{ij}}{\partial x_i} + pf_j$$

Here we used summation convention. The above equation is in fact three equations. We don't have space to explain all the derivation details and the summation conventions. It's important to understand the physical meanings of these equations. The constitutive relations relate the stress tensor with the pressure and fluid velocity field by introducing some material coefficients. Assuming Newtonian fluid, we have:

$$\sigma_{ij} = -p\delta_{ij} + \lambda\delta_{ij}\frac{\partial u_k}{\partial x_k} + \mu\left(\frac{\partial u_i}{\partial x_k} + \frac{\partial u_j}{\partial x_i}\right),$$

where p is pressure, μ is dynamic viscocity, λ is another material coefficient.

Taking this relation into the Equation of momentum conservation, we derive the famous Navier-Stokes Equation:

$$p\frac{\partial u_j}{\partial t} + pu_k\frac{\partial u_j}{\partial x_k} = -\frac{\partial p}{\partial x_j} + \frac{\partial}{\partial x_j}\left(\lambda\frac{\partial u_k}{\partial x_k}\right) + \frac{\partial}{\partial x_i}\left[\mu\left(\frac{\partial u_i}{\partial x_j} + \frac{\partial u_j}{\partial x_i}\right)\right] + f_j p$$

Again this equation uses summation convention and represent three scalar equations.

Under the condition of incompressible flow and the dynamic viscosity to be constant, the Navier-Stokes equation becomes:

$$p\frac{\partial u_j}{\partial t} + pu_k\frac{\partial u_j}{\partial x_k} = -\frac{\partial p}{\partial x_j} + \mu\frac{\partial^2 u_j}{\partial x_i \partial x_i} + f_j p$$

When viscous is neglected, we have the Euler equations:

$$p\frac{\partial u_j}{\partial t} + pu_k\frac{\partial u_j}{\partial x_k} = -\frac{\partial p}{\partial x_j} + f_j p$$

CONSERVATION OF ENERGY

Apply the first law of thermodynamics to fluid element, we have the principle of conservation of energy in fluid mechanics. Consider any arbitrary mass of fluid of volume V and follow it in a lagrangian frame of reference as it flows. The total energy of this mass per unit volume is the combination of internal energy and kinetic energy,

$$pe + \frac{1}{2} p\vec{\mu} \bullet \vec{\mu}\,.$$

The rate of change of the energy of the mass we are following is balanced by the works done by external forces minus the energy the mass give to the outside. In integral form, this relation is:

$$\frac{D}{Dt}\int_V \left(pe + \frac{1}{2} p\vec{\mu} \bullet \vec{\mu} \right) dV = \int_S \vec{\mu} \bullet \vec{P} dS + \int_V \vec{\mu} \bullet p\vec{f} dS - \int_s \vec{q} \bullet \vec{n} dS$$

Applying the relation of continuity and momentum, we get the equation of conservation of thermal energy:

$$p\frac{\partial e}{\partial t} + p\mu_k \frac{\partial e}{\partial x_k} = \sigma_{ij}\frac{\partial u_j}{\partial x_i} - \frac{\partial q_j}{\partial x_j}$$

This equation is the balance of thermal energy only, it comes from the subtraction of mechanical energy from the original energy balance relation. But it is usually refereed to as the Energy Equation.

Taking stress relation into the above equation, we have:

$$p\frac{\partial e}{\partial t} + p\mu_k \frac{\partial e}{\partial x_k} = -p\frac{\partial u_k}{\partial x_k} + \frac{\partial}{\partial x_j}\left(k\frac{\partial T}{\partial x_j} \right) + \Phi$$

where T is temperature, k is the thermal conductivity, Φ is the dissipation function defined as:

$$\Phi = \lambda \left(\frac{\partial u_k}{\partial x_k} \right)^2 + \mu \left(\frac{\partial u_i}{\partial x_j} + \frac{\partial u_j}{\partial x_i} \right) \frac{\partial u_j}{\partial x_i}$$

For incompressible fluid in Cartesian frame,

$$\Phi = \mu \left(\frac{\partial u_i}{\partial x_j} + \frac{\partial u_j}{\partial x_i} \right)^2 .$$

State Equations of Materials

Two state equations are needed to make the above relations self-contained.

$$P = (p, T)$$
$$e = e(p, T)$$

We know for perfect gas P =ρRT, and internal energy e = C_vT, with C_v be the constant volume specific heat.

Some times enthalpy is used in the energy equation, enthalpy is defined as:

$$h = e + P / p = CpT$$

From discussion of 2, 3, 4 and 5, we have shown seven scalar equations with 7 variables (assume k, λ and μ are known). For most general case, these relations are coupled. But usually we can simplify them, either for 1D or 2D flow, or we assume some specific conditions. Assume incompressible flow or neglecting viscous can simplify the equation. If we neglect the velocity in the energy equation, we can get the heat conduction equation. Assume irrotational flow results in the analysis of potential flow.

The model may becomes more complex when turbulence is considered. Turbulence surely takes effect in actual applications, but we don't have time to cover it here. Another complexity comes from shock-wave which is common for high energy laser machining. We will see some discussions in following sections. Are you bored of these mess of math? I guess you need to figure them out in order to read the literatures in this area. I can not take the task of fluid mechanics, if you could not understand the meaning of these relations, you are suggested to turn for a book on Fluid Mechanics. If you feel comfortable for this section, then you will feel comfortable for the following sections. Boundary layers and potential flows are discussed extensively in fluid mechanics. Let's show you some pictures in addition to the boring equations.

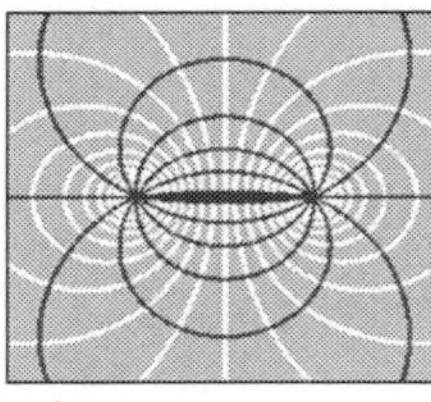

The flow pattern of a dipole

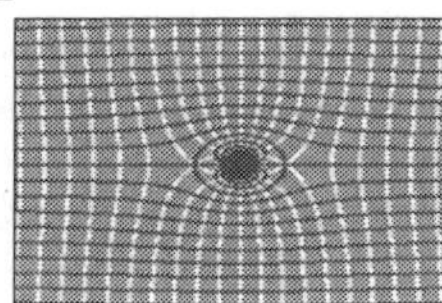

OVERALL HEAT TRANSFER COEFFICIENT

It is often useful to determine values for overall heat transfer coefficients while performing non-exact activities such as early project costestimating and basic heat exchanger performance assessments. The equation which relates the overall heat transfer coefficient to the heat duty and the heat transfer area is:

$$Q = U*A*\Delta Tlm$$

Where:

Q = heat load

U = overall heat transfer coefficient

A = heat transfer area

ΔT_{lm} = log mean temperature difference

Overall heat transfer coefficients are dependant on many parameters such as the nature of the fluid, fluid velocities, type of heat exchanger,temperatures and fouling. Despite all these determining parameters, typical overall heat transfer coefficients are available for common applications and fluids. If little information about the process and the parameters outlined above is available, the following values can be used as a guide for overall heat transfer coefficients:

Sensible Vapour: 30 Btu/hr-ft^2-F

Sensible Heating/Cooling or Condensing: 100 Btu/hr-ft^2-F

Boiling: 120 Btu/hr-ft^2-F

When more information about the fluids and process is available, one can use the overall heat transfer coefficient values in the tables below as a guide as to the order of magnitude. Actual overall heat transfer coefficients may be smaller or larger than the values listed.

Heaters (no phase change)		
Hot Fluid	**Cold Fluid**	**Overall U (BTU/hr-ft^2-F)**
Steam	Air	10 – 20
Steam	Water	250 – 750
Steam	Methanol	200 – 700
Steam	Ammonia	200 – 700
Steam	Aqueous solutions	100 – 700
Steam	Light hydrocarbons (viscosity < 0.5 cP)	100 – 200
Steam	Medium hydrocarbons (0.5 cP < viscosity < 1 cP)	50 – 100
Steam	Heavy hydrocarbons (viscosity > 1)	6 – 60
Steam	Gases	5 – 50
Dowtherm	Gases	4 – 40
Dowtherm	Heavy oils	8 – 60
Flue gas	Aromatic hydrocarbon and steam	5 – 10

Evaporators		
Hot Fluid	**Cold Fluid**	**Overall U (BTU/hr-ft^2-F)**
Steam	Water	350 – 750
Steam	Organic solvents	100 – 200
Steam	Light oils	80 – 180
Steam	Heavy oils (vacuum)	25 – 75
Water	Refrigerant	75 – 150
Organic solvents	Refrigerant	30 – 100

Coolers (no phase change)		
Cold Fluid	**Hot Fluid**	**Overall U (BTU/hr-ft^2-F)**
Water	Water	150 – 300
Water	Organic solvent	50 – 150
Water	Gases	3 – 50
Water	Light oils	60 – 160
Water	Heavy oils	10 – 50
Light oil	Organic solvent	20 – 70
Brine	Water	100 – 200
Brine	Organic solvent	30 – 90
Brine	Gases	3 – 50
Organic solvents	Organic solvents	20 – 60
Heavy oils	Heavy oils	8 – 50

Condensers		
Cold Fluid	**Hot Fluid**	**Overall U (BTU/hr-ft^2-F)**
Water	Steam (pressure)	350 -750
Water	Steam (vacuum)	300 – 600
Water or brine	Organic solvent (saturated, atmospheric)	100 – 200
Water or brine	Organic solvent (atmospheric, high non-condensables)	20 – 80
Water or brine	Organic solvent (saturated, vacuum)	50 – 120
Water or brine	Organic solvent (vacuum, high non-condensables)	10 – 50
Water or brine	Aromatic vapours (atmospheric with non-condensables)	5 – 30
Water	Low boiling hydrocarbon(atmospheric)	80 – 200
Water	High boiling hydrocarbon(vacuum)	10 – 30

When the process is well defined, one can use film heat transfer coefficients to calculate the overall heat transfer coefficient.

The overall heat transfer coefficient can be calculated from the film coefficients using the equation:

$$\frac{1}{U} = \frac{1}{h_{out}} + R_{out} + R_{wo} + R_{i0} + \frac{1}{h_{i0}}$$

Where:

U = overall heat transfer coefficient

h_{out} = film coefficient on outside surface

R_{out} = resistance due to fouling on outside surface

R_{wo} = resistance due to metal wall of heat transfer area (corrected to the outside)

R_{io} = resistance due to fouling on inside surface (corrected to the outside)

h_{io} = = film coefficient on inside surface (corrected to the outside)

In order to use the equation above, values for the film heat transfer coefficients must be determined. Film coefficients, just like overall coefficients, are influenced by many parameters such as nature of the fluid, type of heat exchanger, fluid velocity, transport properties and temperature. The tables below provide examples of film coefficients values for various applications. Again, these should be used as a guide as to the order of magnitude and the actual film coefficients may be smaller or larger than the values listed.

No phase change	
Fluid	**Film Coefficient (BTU/hr-ft^2-F)**
Water	300 – 2000
Gases	3 – 50
Organic Solvents	60 – 500
Oils	10 – 120

Condensing	
Fluid	**Film Coefficient (BTU/hr-ft^2-F)**
Steam	1000 – 3000
Organic Solvents	150 – 500
Light Oils	200 – 400
Heavy Oils (vacuum)	20 – 50
Ammonia	500 – 1000

Evaporation	
Fluid	**Film Coefficient (BTU/hr-ft^2-F)**
Water	800 – 2000
Organic Solvents	100 – 300
Light Oils	150 – 300
Heavy Oils	10 – 50
Ammonia	200 – 400

HEAT FLOW THROUGH WALL WITH VARIABLE THERMAL CONDUCTIVITY

Fourier's law of heat conduction, coefficient of thermal conductivity, effect of temperature and pressure on thermal conductivity of solids, liquids and gases and its measurement.

Three-dimensional general conduction equation in rectangular, cylindrical and spherical coordinates involving internal heat generation and unsteady state conditions. Derivation of equations for simple one dimensional steady state heat conduction from three dimensional equations for heat conduction though walls, cylinders and spherical shells (simple and composite), electrical analogy of the heat transfer phenomenon in the cases discussed above.

Equivalent areas, shape factor, conduction through edges and corners of walls and critical thickness of insulation layers on electric wires and pipes carrying hotfiuids. Internal generation cases along with some practical cases of heat conduction like heat transfer through underground electrical cables, simple model of heat conduction through piston crown and case of nuclear fuel rod with cladding. Influence of variable thermal conductivity on conduction through simple cases of walls I cylinders and spheres. Introduction to unsteady heat transfer, Newtonian heating and cooling of solids; definition and cxplanation of the term thermal diffusivity.

Process of heat flow, definition of emissivity, absorptivity, reflectivity and trarismissMty. Concept of black arid grey bodies, Plank's law of non-chromatic radiation. Kirchoff's law and Stefan Boltzmann law. Interchange factor. Lambert's Cosine law and the geometric factor. Intensity of Radiation (Definition only), radiation density, irradiation, radiosity and radiation shields..

Derivation formula for radiation exchange between two bodies using the definition of radiosity and irradiation and its application to cases of radiation exchange between three or four bodies (*e.g.* boiler or other furnaces), simplification of the formula for its application to simple bodies like two parallel surfaces, concentric cylinders and a body enveloped by an other body etc. Error in Temperature measurement by a thermocouple probe due to radiation losses.

CRITICAL THICKNESS OF INSULATION

We know that by adding more insulation to a wall always decreases heat transfer. The thicker the insulation, the lower the heat transfer rate. This is expected, since the heat transfer area A is constant, and adding insulation always increases the thermal resistance of the wall without affecting the convection resistance.

Adding insulation to a cylindrical piece or a spherical shell, however, is a different matter. The additional insulation increases the conduction resistance of the insulation layer but decreases the convection resistance of the surface because of the increase in the outer surface area for convection. The heat transfer from the pipe may increase or decrease, depending on which effect dominates. Consider a cylindrical pipe of outer radius r_1 whose outer surface temperature T_1 is maintained constant (Figure.).

The pipe is now insulated with a material whose thermal conductivity is k and outer radius is r_2.

Heat is lost from the pipe to the surrounding medium at temperature T_∞, with a convection heat transfer coefficient h. The rate of heat transfer from the insulated pipe to the surrounding air can be expressed as (Figure.)

$$q_r = \frac{(T_1 - T_\infty)}{\dfrac{ln\left(r_2/r_1\right)}{2\pi Lk} + \dfrac{1}{h(2\pi r_2 L)}}$$

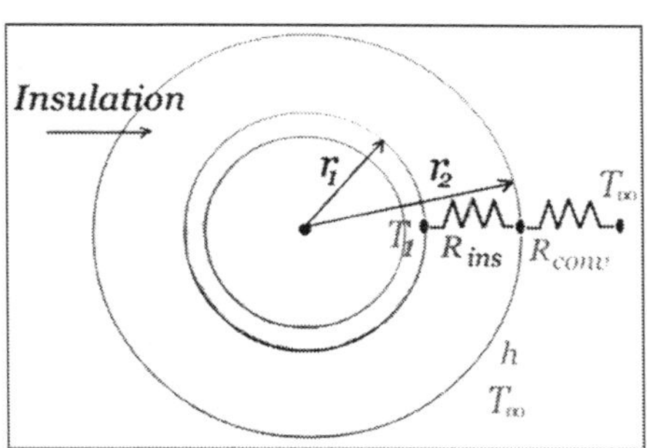

Fig. Insulated Cylindrical Pipe

The variation of heat transfer rate with the outer radius of insulation r_2 is plotted in Figure. The value of r_2 at which heat transfer rate reaches maximum is determined from the requirement that,

$$\frac{dq_r}{dr} \text{ (zero slope).}$$

Performing the differentiation and solving for r_2 yields the critical radius of insulation for a cylindrical body to be,

$$r_{cr,cylinder} = \frac{k}{h}$$

Note that the critical radius of insulation depends on the thermal conductivity of the insulation k and the external convection heat transfer coefficient h. The rate of heat transfer from the cylinder increases with the addition of insulation for $r_2 < r_{cr}$, reaches a maximum when $r_2 = r_{cr}$, and starts to decrease for $r_2 > r_{cr}$.

Thus, insulating the pipe may actually increase the rate of heat transfer from the pipe instead of decreasing it when $r_2 < r_{cr}$.

The important question to answer at this point is to whether we need to be concerned about the critical radius of insulation when insulating hot water pipes or even hot water tanks. Should we always check and make sure that the outer radius of insulation exceeds the critical radius before we install any insulation? Probably not, as explained below.

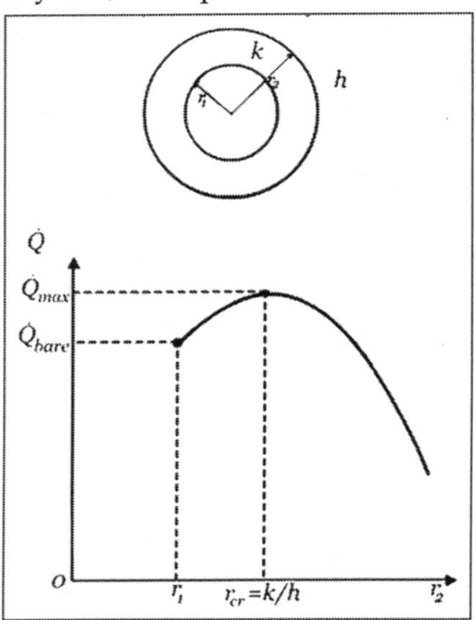

Fig. Variation of Heat Transfer Rate With Radius

The value of the critical radius rcr will be the largest when k is large and h is small. Noting that the lowest value ofh encountered in practice is about 5 W/m2K for the case of natural convection of gases, and that the thermal conductivity of common insulating materials is 0.05 W/m2K, the largest value of the critical radius we are likely to encounter is,

$$r_{cr} = \frac{k_{\max,insulation}}{h_{\min}} = \frac{0.05}{5} = 0.01m = 10mm$$

This value would be even smaller when the radiation effects are considered. The critical radius would be much less in forced convection, often less than 1 mm, because of much larger h values associated with forced convection. Therefore, we can insulate hot water or steam pipes freely without worrying about the possibility of increasing the heat transfer by insulating the pipes.

The radius of electric wires may be smaller than the critical radius. Therefore, the plastic electrical insulation may actually enhance the heat transfer from electric wires and thus keep their steady operating temperatures at lower and thus safer levels.

The discussions above can be repeated for a sphere, and it can be shown in a similar manner that the critical radius of insulation for a spherical shell is,

$$r_{cr,sphere} = \frac{2k}{h}$$

where k is the thermal conductivity of the insulation and h is the convection heat transfer coefficient on the outer surface.

3

Mechanisms of Heat Transfer

Heat transfer is a discipline of thermal engineering that concerns the generation, use, conversion, and exchange of thermal energy and heat between physical systems. Heat transfer is classified into various mechanisms, such as heat conduction, convection, thermal radiation, and transfer of energy by phase changes. Engineers also consider the transfer of mass of differing chemical species, either cold or hot, to achieve heat transfer. While these mechanisms have distinct characteristics, they often occur simultaneously in the same system.

Heat conduction, also called diffusion, is the direct microscopic exchange of kinetic energy of particles through the boundary between two systems. When an object is at a different temperature from another body or its surroundings, heat flows so that the body and the surroundings reach the same temperature, at which point they are in thermal equilibrium. Such spontaneous heat transfer always occurs from a region of high temperature to another region of lower temperature, as required by the second law of thermodynamics.

Heat convection occurs when bulk flow of a fluid (gas or liquid) carries heat along with the flow of matter in the fluid. The flow of fluid may be forced by external processes, or sometimes (in gravitational fields) by buoyancy forces caused when thermal energy expands the fluid (for example in a fire plume), thus influencing its own transfer. The latter process is often called "natural convection". All convective processes also move heat partly by diffusion, as well. Another form of convection is forced convection. In this case the fluid is forced to flow by use of a pump, fan or other mechanical means.

The final major form of heat transfer is by radiation, which occurs in any transparent medium (solid or fluid) but may also even occur across vacuum (as when the Sun heats the Earth). Radiation is the transfer of energy through space by means of electromagnetic waves in much the same way as electromagnetic light waves transfer light. The same laws that govern the transfer of light govern the radiant transfer of heat.

Heat is defined in physics as the transfer of thermal energy across a well-defined boundary around a thermodynamic system. It is a characteristic of a process and is not statically contained in matter. In engineering contexts, however, the term *heat transfer* has acquired a specific usage, despite its literal redundancy of the characterization of transfer. In these contexts, *heat* is taken as synonymous to thermal energy. This usage has its origin in the historical interpretation of heat as a fluid (*caloric*) that can be transferred by various causes, and that is also common in the language of laymen and everyday life.

Fundamental methods of heat transfer in engineering include conduction, convection, and radiation. Physical laws describe the behaviour and characteristics of each of these methods. Real systems often exhibit a complicated combination of them. Heat transfer methods are used in numerous disciplines, such as automotive engineering, thermal management of electronic devices and systems, climate control, insulation, materials processing, and power plant engineering.

Various mathematical methods have been developed to solve or approximate the results of heat transfer in systems. Heat transfer is a path function (or process quantity), as opposed to a state quantity; therefore, the amount of heat transferred in a thermodynamic process that changes the state of a system depends on how that process occurs, not only the net difference between the initial and final states of the process. Heat flux is a quantitative, vectorial representation of the heat flow through a surface.

Heat transfer is typically studied as part of a general chemical engineering or mechanical engineering curriculum. Typically, thermodynamics is a prerequisite for heat transfer courses, as the laws of thermodynamics are essential to the mechanism of heat transfer. Other courses related to heat transfer include energy conversion, thermofluids, and mass transfer.

The transport equations for thermal energy (Fourier's law), mechanical momentum (Newton's law for fluids), and mass transfer (Fick's laws of diffusion) are similar and analogies among these three transport processes have been developed to facilitate prediction of conversion from any one to the others.

HEAT TRANSFER IN THE HUMAN BODY

The principles of heat transfer in engineering systems can be applied to the human body in order to determine how the body transfers heat. Heat is produced in the body by the continuous metabolism of nutrients which provides energy for the systems of the body. The human body must maintain a consistent internal temperature in order to maintain healthy bodily functions. Therefore, excess heat must be dissipated from the body to keep it from overheating. When a person engages in elevated levels of physical activity, the body requires additional fuel which increases the metabolic rate and the rate of heat production. The body must then use additional methods to remove the additional heat produced in order to keep the internal temperature at a healthy level.

Heat transfer by convection is driven by the movement of fluids over the surface of the body. This convective fluid can be either a liquid or a gas. For heat transfer from the outer surface of the body, the convection mechanism is dependent on the surface area of the body, the velocity of the air, and the temperature gradient between the surface of the skin and the ambient air. The normal temperature of the body is approximately 37 °C. Heat transfer occurs more readily when the temperature of the surroundings is significantly less than the normal body temperature.

This concept explains why a person feels "cold" when not enough covering is worn when exposed to a cold environment. Clothing can be considered an insulator which provides thermal resistance to heat flow over the covered portion of the body. This thermal resistance causes the temperature on the surface of the clothing to be less than the temperature on the surface of the skin. This smaller temperature gradient between the surface temperature and the ambient temperature will cause a lower rate of heat transfer than if the skin were not covered.

In order to ensure that one portion of the body is not significantly hotter than another portion, heat must be distributed evenly through the bodily tissues. Blood flowing through blood vessels acts as a convective fluid and helps to prevent any buildup of excess heat inside the tissues of the body. This flow of blood through the vessels can be modeled as pipe flow in an engineering system. The heat carried by the blood is determined by the temperature of the surrounding tissue, the diameter of the blood vessel, the thickness of the fluid, velocity of the flow, and the heat transfer coefficient of the blood. The velocity, blood vessel diameter, and the fluid thickness can all be related with the Reynolds Number, a dimensionless number used in fluid mechanics to characterize the flow of fluids.

Latent heat loss, also known as evaporative heat loss, accounts for a large fraction of heat loss from the body. When the core temperature of the body increases, the body triggers sweat glands in the skin to bring additional moisture to the surface of the skin. The liquid is then transformed into vapour which removes heat from the surface of the body. The rate of evaporation heat loss is directly related to the vapour pressure at the skin surface and the amount of moisture present on the skin. Therefore, the maximum of heat transfer will occur when the skin is completely wet. The body continuously loses water by evaporation but the most significant amount of heat loss occurs during periods of increased physical activity.

heat pipe

A heat pipe is a passive device constructed in such a way that it acts as though it has extremely high thermal conductivity. Heat pipes use latent heat and capillary action to move heat, and can carry many times as much heat as a similar-sized copper rod. Originally invented for use in satellites, they have applications in personal computers. A thermocouple is a junction between two different metals that produces a voltage related to a temperature difference. Thermocouples are a widely used type of temperature sensor for measurement and control, and can also be used to convert heat into electric power.

A thermopile is an electronic device that converts thermal energy into electrical energy. It is composed of thermocouples. Thermopiles do not measure the absolute temperature, but generate an output voltage proportional to a temperature difference. Thermopiles are widely used, *e.g.*, they are the key component of infrared thermometers, such as those used to measure body temperature via the ear. A thermal diode or thermal rectifier is a device that preferentially passes heat in one direction: a "one-way valve" for heat.

FUNDAMENTAL MODES OF HEAT TRANSFER

MECHANISMS

The fundamental modes of heat transfer are:

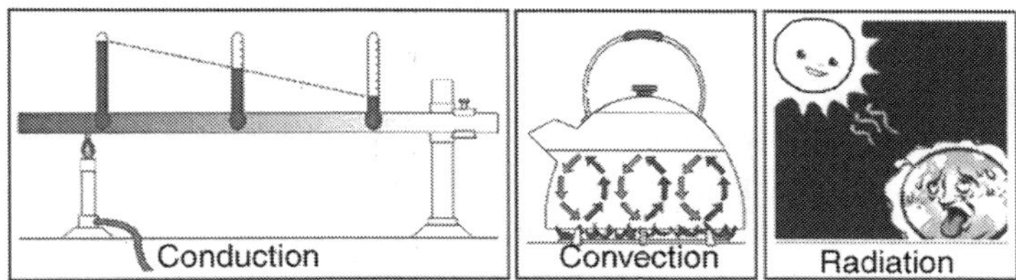

- *Conduction or diffusion*: The transfer of energy between objects that are in physical contact
- *Convection*: The transfer of energy between an object and its environment, due to fluid motion
- *Radiation*: The transfer of energy to or from a body by means of the emission or absorption of electromagnetic radiation
- *Advection*: The transfer of energy from one location to another as a side effect of physically moving an object containing that energy

CONDUCTION OF HEAT TRANSFER

In heat transfer, conduction (or heat conduction) is a mode of transfer of energy within and between bodies of matter, due to a temperature gradient. Conduction means collisional and diffusive transfer of kinetic energy through particles of ponderable matter (as distinct from photons). Conduction takes place in all forms of ponderable matter, such as solids, liquids, gases and plasmas. Heat spontaneously tends to flow from a body at a higher temperature to a body at a lower temperature. In the absence of external driving fluxes, temperature differences, over time, approach thermal equilibrium. In conduction, the heat flows through the body itself, as opposed to its transfer by the bulk motion of the matter as in convection, and by thermal radiation. In solids, it is due to the combination of vibrations of the molecules in a lattice or phonons with the energy transported by free electrons. In gases and liquids, conduction is due to the collisions and diffusion of the molecules during their random motion. Photons in general do not collide with one another and thermal transport by electromagnetic radiation is not regarded as conduction of heat. In solids, it is not simple to separate transfer by photons from transfer by ponderable matter, but the distinction can be more easily made in liquids, and is routinely made in gases.

In the engineering sciences, heat transfer includes the processes of thermal radiation, convection, and sometimes mass transfer. Usually more than one of these processes occurs in a given situation. The conventional symbol for the material property, thermal conductivity, is k. On a microscopic scale, conduction occurs as rapidly moving or vibrating atoms and molecules interact with neighboring particles, transferring some of their kinetic energy. Heat is transferred by conduction when adjacent atoms vibrate against one another, or as electrons move from one atom to another. Conduction is the most significant means of heat transfer within a solid or between solid objects in thermal contact. Conduction is greater in solids because the network of relatively fixed spatial relationships between atoms helps to transfer energy between them by vibration.

Fluids (and especially gases) are less conductive. This is due to the large distance between atoms in a gas: fewer collisions between atoms means less conduction. Conductivity of gases increases with temperature. Conductivity increases with increasing pressure from vacuum up to a critical point that the density of the gas is such that molecules of the gas may be expected to collide with each other before they transfer heat from one surface to another. After this point conductivity increases only slightly with increasing pressure and density. Thermal contact conductance is the study of heat conduction between solid bodies in contact.

A temperature drop is often observed at the interface between the two surfaces in contact. This phenomenon is said to be a result of a thermal contact resistance existing between the contacting surfaces. Interfacial thermal resistance is a measure of an interface's resistance to thermal flow. This thermal resistance differs from contact resistance, as it exists even at atomically perfect interfaces. Understanding the thermal resistance at the interface between two materials is of primary significance in the study of its thermal properties. Interfaces often contribute significantly to the observed properties of the materials.

The inter-molecular transfer of energy could be primarily by elastic impact as in fluids or by free electron diffusion as in metals or phonon vibration as in insulators. In insulators the heat flux is carried almost entirely by phonon vibrations. Metals (*e.g.* copper, platinum, gold,etc.) are usually good conductors of thermal energy. This is due to the way that metals are chemically bonded: metallic bonds (as opposed to covalent or ionic bonds) have free-moving electrons which are able to transfer thermal energy rapidly through the metal.

The "electron fluid" of a conductive metallic solid conducts most of the heat flux through the solid. Phonon flux is still present, but carries less of the energy. Electrons also conduct electric current through conductive solids, and the thermal and electrical conductivities of most metals have about the same ratio. A good electrical conductor, such as copper, also conducts heat well. Thermoelectricity is caused by the interaction of heat flux and electrical current. Heat conduction within a solid is directly analogous to diffusion of particles within a fluid, in the situation where there are no fluid currents.

To quantify the ease with which a particular medium conducts, engineers employ the thermal conductivity, also known as the conductivity constant or conduction coefficient, *k*. In thermal conductivity *k* is defined as "the quantity of heat, Q, transmitted in time (t) through a thickness (L), in a direction normal to a surface of area (A), due to a temperature difference (DT) [...]." Thermal conductivity is a material *property* that is primarily dependent on the medium's phase, temperature, density, and molecular bonding. Thermal effusivity is a quantity derived from conductivity which is a measure of its ability to exchange thermal energy with its surroundings.

FORM OF CONDUCTION

Steady-state Conduction

Steady state conduction is the form of conduction that happens when the temperature difference(s) driving the conduction are constant, so that (after an equilibration time), the spatial distribution of temperatures (temperature field) in the conducting object does not change any further. Thus, all partial derivatives of temperature *with respect to space* may either be zero or have nonzero values, but all derivatives of temperature at any point *with respect to time* are uniformly zero. In steady state conduction, the amount of heat entering any region of an object is equal to amount of heat coming out (if this were not so, the temperature would be rising or falling, as thermal energy was tapped or trapped in a region).

For example, a bar may be cold at one end and hot at the other, but after a state of steady state conduction is reached, the spatial gradient of temperatures along the bar does not change any further, as time proceeds. Instead, the temperature at any given section of the rod remains constant, and this temperature varies linearly in space, along the direction of heat transfer. In steady state conduction, all the laws of direct current electrical conduction can be applied to "heat currents". In such cases, it is possible to take "thermal resistances" as the analog to electrical resistances. In such cases, temperature plays the role of voltage, and heat transferred per unit time (heat power) is the analog of electrical current.

Steady state systems can be modelled by networks of such thermal resistances in series and in parallel, in exact analogy to electrical networks of resistors. See purely resistive thermal circuits for an example of such a network.

PERTURBATION OF TEMPERATURE

Transient Conduction

In general, during any period in which temperatures are changing *in time* at any place within an object, the mode of thermal energy flow is termed *transient conduction.* Another term is "non steady-state" conduction, referring to time-dependence of temperature fields in an object. Non-steady-state situations appear after an imposed change in temperature at a boundary of an object. They may also occur with temperature changes inside an object, as a result of a new source or sink of heat suddenly introduced within an object, causing temperatures near the source or sink to change in time.

When a new perturbation of temperature of this type happens, temperatures within the system will change in time towards a new equilibrium with the new conditions, provided that these do not change. After equilibrium, heat flow into the system will once again equal the heat flow out, and temperatures at each point inside the system no longer change. Once this happens, transient conduction is ended, although steady-state conduction may continue if there continues to be heat flow.

If changes in external temperatures or internal heat generation changes are too rapid for equilibrium of temperatures in space to take place, then the system never reaches a state of unchaging temperature distribution in time, and the system remains in a transient state. An example of a new source of heat "turning on" within an object which causes transient conduction, is an engine starting in an automobile. In this case the transient thermal conduction phase for the entire machine would be over, and the steady state phase would appear, as soon as the engine had reached steady-state operating temperature. In this state of steady-state equilibrium, temperatures would vary greatly from the engine cylinders to other parts of the automobile, but at no point in space within the automobile would temperature be increasing or decreasing. After establishment of this state, the transient conduction phase of heat transfer would be over.

New external conditions also cause this process: For example the copper bar in the example steady-state conduction would experience transient conduction as soon as one end was subjected to a different temperature from the other. Over time, the field of temperatures inside the bar would reach a new steady-state, in which a constant temperature gradient along the bar will finally be set up, and this gradient would then stay constant in space. Typically, such a new steady state gradient is approached exponentially with time after a new temperature-or-heat source or sink, has been introduced. When a "transient conduction" phase is over, heat flow may still continue at high power, so long as temperatures do not change.

An example of transient conduction which does not end with steady-state conduction, but rather no conduction, occurs when a hot copper ball is dropped into oil at a low temperature. Here the temperature field within the object begins to change as a function of time, as the heat is removed from the metal, and the interest lies in analysing this spatial change of temperature within the object over time, until all gradients disappear entirely (the ball has reached the same temperature as the oil). Mathematically, this condition is also approached exponentially; in theory it takes infinite time, but in practice it is over, for all intents and purposes, in a much shorter period. At the end of this process with no heat sink but the internal parts of the ball (which are finite), there is no steady state heat conduction to be reached. Such a state never occurs in this situation, but rather the end of the process is when there is no heat conduction at all.

Analysis of non steady-state conduction systems is more complex than steady-state systems, and (except for simple shapes) calls for the application of approximation theories, and/or numerical analysis by computer. One popular graphical method involves the use of Heisler Charts. Occasionally transient conduction problems may be considerably simplified if regions of the object being heated or cooled can be identified, in which thermal conductivity is very much greater than that for heat paths leading into the region. In this case, the region with high conductivity can often be treated in the lumped capacitance model, as a "lump" of material with a simple thermal capacitance consisting of its aggregate heat capacity.

Such regions show no temperature variation across their extent during warming or cooling (as compared to the rest of the system) due to their far higher conductance. During transient conduction, therefore, their temperature changes uniformly in space, and as a simple exponential in time. An example of such systems are those which follow "Newton's law of cooling" during transient cooling (or the reverse during heating). The equivalent thermal circuit consists of a simple capacitor in series with a resistor. In such cases, the remainder of the system with high thermal resistance (comparatively low conductivity) plays the role of the resistor in the circuit.

THEORY OF RELATIVISTIC HEAT CONDUCTION

RELATIVISTIC CONDUCTION

The theory of relativistic heat conduction is a model that is compatible with the theory of special relativity. For most of the last century, it was recognized that Fourier equation is in contradiction with the theory of relativity because it admits an infinite speed of propagation of heat signals. For example, according to Fourier equation, a pulse of heat at the origin would be felt at infinity instantaneously.

The speed of information propagation is faster than the speed of light in vacuum, which is physically inadmissible within the framework of relativity. Alterations to the Fourier model provided for a relativistic model of heat conduction, avoiding this problem.

Quantum Conduction

Second sound is a quantum mechanical phenomenon in which heat transfer occurs by wave-like motion, rather than by the more usual mechanism of diffusion. Heat takes the place of pressure in normal sound waves. This leads to a very high thermal conductivity. It is known as "second sound" because the wave motion of heat is similar to the propagation of sound in air.

Fourier's Law

The law of heat conduction, also known as Fourier's law, states that the time rate of heat transfer through a material is proportional to the negative gradient in the temperature and to the area, at right angles to that gradient, through which the heat is flowing. We can state this law in two equivalent forms: the integral form, in which we look at the amount of energy flowing into or out of a body as a whole, and the differential form, in which we look at the flow rates or fluxes of energy locally.

Newton's law of cooling is a discrete analog of Fourier's law, while Ohm's law is the electrical analogue of Fourier's law.

Differential Form

The differential form of Fourier's Law of thermal conduction shows that the local heat flux density, $\vec{q}$, is equal to the product of thermal conductivity, k, and the negative local temperature gradient, $-\nabla T$. The heat flux density is the amount of energy that flows through a unit area per unit time.

$$\vec{q} = -k\nabla T$$

where (including the SI units) $\vec{q}$

is the local heat flux, $W \cdot m^{-2}$

k is the material's conductivity, $W \cdot m^{-1} \cdot K^{-1}$,

∇T is the temperature gradient, $K \cdot m^{-1}$.

The thermal conductivity, k, is often treated as a constant, though this is not always true. While the thermal conductivity of a material generally varies with temperature, the variation can be small over a significant range of temperatures for some common materials. In anisotropic materials, the thermal conductivity typically varies with orientation; in this case k is represented by a second-order tensor. In nonuniform materials, k varies with spatial location.

For many simple applications, Fourier's law is used in its one-dimensional form. In the x-direction,

$$q_x = -k\frac{dT}{dx}$$

Integral Form

By integrating the differential form over the material's total surface S, we arrive at the integral form of Fourier's law:

$$\frac{\partial Q}{\partial t} = -k\oint_S \vec{\nabla} T \cdot \overrightarrow{dA}$$

where (including the SI units)

: $\frac{\partial Q}{\partial t}$ is the amount of heat transferred per unit time (in W) and

: $\overrightarrow{dA}$ is an oriented surface area element (in m^2)

The above differential equation, when integrated for a homogeneous material of 1-D geometry between two endpoints at constant temperature, gives the heat flow rate as:

$$\frac{\Delta Q}{\Delta t} = -kA\frac{\Delta T}{\Delta x}$$

where,

A is the cross-sectional surface area,

ΔT is the temperature difference between the ends,

Δx is the distance between the ends.

This law forms the basis for the derivation of the heat equation.

Conductance

Writing,

$$U = \frac{k}{\Delta x},$$

where *U* is the conductance, in W/(m^2 K).

Fourier's law can also be stated as:

$$\frac{\Delta Q}{\Delta t} = UA(-\Delta T).$$

The reciprocal of conductance is resistance, R, given by:

$$R = \frac{1}{U} = \frac{\Delta x}{k} = \frac{(-\Delta T)}{\frac{\Delta Q}{\Delta t}}$$

and it is resistance which is additive when several conducting layers lie between the hot and cool regions, because *A* and *Q* are the same for all layers. In a multilayer partition, the total conductance is related to the conductance of its layers by:

$$\frac{1}{U} = \frac{1}{U_1} + \frac{1}{U_2} + \frac{1}{U_3} + \cdots$$

So, when dealing with a multilayer partition, the following formula is usually used:

$$\frac{\Delta Q}{\Delta t} = \frac{A(-\Delta T)}{\frac{D_{x1}}{k_1} + \frac{D_{x2}}{k_2} + \frac{D_{x3}}{k_3} + \cdots}.$$

When heat is being conducted from one fluid to another through a barrier, it is sometimes important to consider the conductance of the thin film of fluid which remains stationary next to the barrier. This thin film of fluid is difficult to quantify, its characteristics depending upon complex conditions of turbulence and viscosity, but when dealing with thin high-conductance barriers it can sometimes be quite significant.

REFORMULATED IN TERMS OF INTENSIVE PROPERTIES

Intensive-property Representation

The previous conductance equations, written in terms of extensive properties, can be reformulated in terms of intensive properties.

Ideally, the formulae for conductance should produce a quantity with dimensions independent of distance, like Ohm's Law for electrical resistance: R = V/I, and conductance: G = I/V.

From the electrical formula: $R = \rho x/A$, where ñ is resistivity, x is length, and A is cross-sectional area, we have $G = kA/x$, where G is conductance, k is conductivity, x is length, and A is cross-sectional area.

For Heat,

$$U = \frac{kA}{\Delta x},$$

where *U* is the conductance.

Fourier's law can also be stated as:

$$h\dot{h}Q = U\,\Delta T$$

analogous to Ohm's law: I = V/R or I = VG.

The reciprocal of conductance is resistance, R, given by:

$$R = \frac{DT}{\dot{Q}},$$

analogous to Ohm's law: R = V/I.

The rules for combining resistances and conductances (in series and in parallel) are the same for both heat flow and electric current.

Cylindrical Shells

Conduction through cylindrical shells (*e.g.* pipes) can be calculated from the internal radius, r_1, the external radius, r_2, the length, l, and the temperature difference between the inner and outer wall, $T_2 - T_1$.

The surface area of the cylinder is $A_r = 2\pi r l$

When Fourier's equation is applied:

$$\dot{Q} = -kA_r \frac{dT}{dr} = -2k\pi r\ell \frac{dT}{dr}$$

and rearranged:

$$\dot{Q}\int_{r_1}^{r_2} \frac{1}{r} dr = -2k\pi\ell \int_{T_1}^{T_2} dT$$

then the rate of heat transfer is:

$$\dot{Q} = 2k\pi\ell \frac{T_1 - T_2}{\text{In}(r_2 / r_1)}$$

the thermal resistance is:

$$R_c = \frac{\Delta T}{\dot{Q}} = \frac{\text{In}(r_2 / r_1)}{2\pi k\ell}$$

and $\dot{Q} = 2\pi k\ell r_m \frac{T_1 - T_2}{r_2 - r_1}$, where $r_m = \frac{r_2 - r_1}{\text{In}(r_2 / r_1)}$. It is important to note that this is the log-mean radius.

Spherica

The conduction through a spherical shell with internal radius, r_1, and external radius, r_2, can be calculated in a similar manner as for a cylindrical shell.

The surface area of the sphere is: $A = 4\pi r^2$

Solving in a similar manner as for a cylindrical shell (see above) produces:

$$\dot{Q} = 4k\pi\frac{T_1 - T_2}{1/r_1 - 1/r_2}$$

ZEROTH LAW OF THERMODYNAMICS

One statement of the so-called zeroth law of thermodynamics is directly focused on the idea of conduction of heat. Bailyn writes that "... the zeroth law may be stated:

All diathermal walls are equivalent."

A diathermal wall is a connection of contiguity between two bodies that allows the passage of heat by conduction between them.

This statement of the 'zeroth law' belongs to an idealized theoretical discourse, and actual physical walls do not match its grandiloquent generality.

But with suitable restrictions, the statement has physical import. For example, the material of the wall must not suffer a phase transition, such as evaporation or fusion, at the temperature at which it has to conduct heat. But when only thermal equilibrium is being considered, and time is not urgent, so that the conductivity of the material does not matter too much, one suitable conductor of heat is as good as another. Conversely, another aspect of the zeroth law is that, subject again to suitable restrictions, a given diathermal wall is indifferent to the nature of the heat bath to which it is connected. For example the glass bulb of a thermometer will act as a diathermal wall whether exposed to a gas or to a liquid, provided they do not corrode it or melt it.

These indifferences are amongst the defining characteristics of heat transfer. In a sense they are symmetries of heat transfer

CONVECTIVE HEAT TRANSFER

CONVECTION

Convective heat transfer, or convection, is the transfer of heat from one place to another by the movement of fluids, a process that is essentially transfer of heat via mass transfer. (In physics, the term *fluid* means any substance that deforms under shear stress; it includes liquids, gases, plasmas, and some plastic solids.) Bulk motion of fluid enhances heat transfer in many physical situations, such as (for example) between a solid surface and the fluid. Convection is usually the dominant form of heat transfer in liquids and gases. Although sometimes discussed as a third method of heat transfer, convection is usually used to describe the combined effects of heat conduction within the fluid (diffusion) and heat transference by bulk fluid flow streaming. The process of transport by fluid streaming is known as advection, but pure advection is a term that is generally associated only with mass transport in fluids, such as advection of pebbles in a river. In the case of heat transfer in fluids, where transport by advection in a fluid is always also accompanied by transport via heat diffusion (also known as heat conduction) the process of heat convection is understood to refer to the sum of heat transport by advection and diffusion/conduction.

Free, or natural, convection occurs when bulk fluid motion (steams and currents) are caused by buoyancy forces that result from density variations due to variations of temperature in the fluid. *Forced* convection is a term used when the streams and currents in the fluid are induced by external means—such as fans, stirrers, and pumps—creating an artificially induced convection current.

Convective heating or cooling in some circumstances may be described by Newton's law of cooling: "The rate of heat loss of a body is proportional to the difference in temperatures between the body and its surroundings." However, by definition, the validity of Newton's law of cooling requires that the rate of heat loss from convection be a linear function of ("proportional to") the temperature difference that drives heat transfer, and in convective cooling this is sometimes not the case. In general, convection is not linearly dependent on temperature gradients, and in some cases is strongly nonlinear. In these cases, Newton's law does not apply.

The term convection can refer to transfer of heat with any fluid movement, but advection is the more precise term for the transfer due only to bulk fluid flow. The process of transfer of heat from a solid to a fluid, or the reverse, requires not only transfer of heat by bulk motion of the fluid, but also diffusion/conduction of heat through the still boundary layer next to the solid. Thus, this process with a moving fluid requires both diffusion and advection of heat, a summed process that is generally called convection.

Convection can be "forced" by movement of a fluid by means other than buoyancy forces (for example, a water pump in an automobile engine). In some cases, natural buoyancy forces alone are entirely responsible for fluid motion when the fluid is heated, and this process is called "natural convection." An example is the draft in a chimney or around any fire. In natural convection, an increase in temperature produces a reduction in density, which causes fluid motion due to pressures and forces when fluids of different densities are affected by gravity (or any g-force). For example, when water is heated on a stove, hot water from the bottom of the pan rises, displacing the colder denser liquid which falls. After heating has stopped, mixing and conduction from this natural convection eventually result in a nearly homogeneous density, and even temperature.

The convection heat transfer mode is comprised to two mechanisms. In addition to energy transfer due to random molecular motion (diffusion), energy is also transferred by bulk, or macroscopic, motion of the fluid. This motion is associated with the fact that, at any instant, large numbers of molecules are moving collectively or as aggregates. Such motion, in the presence of a temperature gradient, contributes to heat transfer. Because the molecules in aggregate retain their random motion, the total heat transfer is then due to the superposition of energy transport by random motion of the molecules and by the bulk motion of the fluid. It is customary to use the term convection when referring to this cumulative transport and the term advection when referring to the transport due to bulk fluid motion.

Convection is the transfer of thermal energy from one place to another by the movement of fluids. Convection is usually the dominant form of heat transfer in liquids and gases. Although often discussed as a distinct method of heat transfer, convection describes the combined effects of conduction and fluid flow or mass exchange.

Two types of convective heat transfer may be distinguished:

1. *Free or natural convection*: When fluid motion is caused by buoyancy forces that result from the density variations due to variations of temperature in the fluid. In the absence of an external source, when the fluid is in contact with a hot surface, its molecules separate and scatter, causing the fluid to be less dense. As a consequence, the fluid is displaced while the cooler fluid gets denser and the fluid sinks. Thus, the hotter volume transfers heat towards the cooler volume of that fluid. Familiar examples are the upward flow of air due to a fire or hot object and the circulation of water in a pot that is heated from below.
2. *Forced convection*: When a fluid is forced to flow over the surface by an external source such as fans, by stirring, and pumps, creating an artificially induced convection current.

Internal and external flow can also classify convection. Internal flow occurs when a fluid is enclosed by a solid boundary such when flowing through a pipe. An external flow occurs when a fluid extends indefinitely without encountering a solid surface. Both of these types of convection, either natural or forced, can be internal or external because they are independent of each other. The bulk temperature, or the average fluid temperature, is a convenient reference point for evaluating properties related to convective heat transfer, particularly in applications related to flow in pipes and ducts.

For a visual experience of natural convection, a glass filled with hot water and some red food dye may be placed inside a fish tank with cold, clear water. The convection currents of the red liquid may be seen to rise and fall in different regions, then eventually settle, illustrating the process as heat gradients are dissipated.

CONVECTION COOLING

NEWTON'S LAW OF COOLING

Convection-cooling can sometimes be described by Newton's law of cooling in cases where the heat transfer coefficient is independent or relatively independent of the temperature difference between object and environment. This is sometimes true, but is not guaranteed to be the case.

Newton's law, which requires a constant heat transfer coefficient, states that *the rate of heat loss of a body is proportional to the difference in temperatures between the body and its surroundings.* The rate of heat transfer in such circumstances is derived below:

Newton's cooling law is a solution of the differential equation given by Fourier's law:

$$\frac{dQ}{dt} = \dot{Q} = h \cdot A \cdot \left(T_{env} - T(t)\right) = -h \cdot A\Delta T(t)$$

where,

Q is the thermal energy in joules

h is the Heat transfer coefficient (assumed independent of T here)

A is the surface area of the heat being transferred

T is the temperature of the object's surface and interior (since these are the same in this approximation)

T_{env} is the temperature of the environment; *i.e.* the temperature suitably far from the surface

DT(t) = T(t) – T_{env} is the time-dependent thermal gradient between environment and object

The heat transfer coefficient h depends upon physical properties of the fluid and the physical situation in which convection occurs. Therefore, a single usable heat transfer coefficient (one that does not vary significantly across the temperature-difference ranges covered during cooling and heating) must be derived or found experimentally for every system analysed. Formulas and correlations are available in many references to calculate heat transfer coefficients for typical configurations and fluids.

For laminar flows, the heat transfer coefficient is rather low compared to the turbulent flows; this is due to turbulent flows having a thinner stagnant fluid film layer on heat transfer surface. However, note that Newton's law breaks down if the flows should transition between laminar or turbulent flow, since this will change the heat transfer coefficient h which is assumed constant in solving the equation.

Newton's law requires that internal heat conduction within the object be large in comparison to the loss/gain of heat by convection (lumped capacitance model), and this may not be true. Also, an accurate formulation for temperatures may require analysis based on changing heat transfer coefficients at different temperatures, a situation frequently found in free-convection situations, and which precludes accurate use of Newton's law. Assuming these are not problems, then the solution can be given if heat transfer within the object is considered to be far more rapid than heat transfer at the boundary (so that there are small thermal gradients within the object). This condition, in turn, allows the heat in the object to be expressed as a simple product of the object's mass, its heat capacity, and its temperature, as in the following section:

Solution in Terms of Object Heat Capacity

If the entire body is treated as lumped capacitance thermal energy reservoir, with a total thermal energy content which is proportional to simple total heat capacity C, and T, the temperature of the body, or Q = CT, it is expected that the system will experience exponential decay with time in the temperature of a body.

From the definition of heat capacity C comes the relation C = dQ/dt. Differentiating this equation with regard to time gives the identity (valid so long as temperatures in the object are uniform at any given time): dQ/dt = C(dT/dt). This expression may be used to replace dQ/dt in the first equation which begins this section, above. Then, if T (t) is the temperature of such a body at time t, and T_{env} is the temperature of the environment around the body:

$$\frac{dT(t)}{dt} = -r\left(T(t) - T_{evn}\right) = -r\Delta T(t)$$

where,

r = hA/C is a positive constant characteristic of the system, which must be in units of S^{-1}, and is therefore sometimes expressed in terms of a characteristic time constant t_0 given by: $r = 1/t_0 = \Delta T/\left(dT(t)/dt\right)$. Thus, in thermal systems, t_0 = C/hA. (The total heat capacity C of a system may be further represented by its mass-specific heat capacity C_p multiplied by its mass m, so that the time constant t_0 is also given by mc_p/hA).

The solution of this differential equation, by standard methods of integration and substitution of boundary conditions, gives:

$$T(t) = T_{env} + \left(T(0) - T_{env}\right)e^{-rt}.$$

If:

ΔT(t) is defined as: T(t) – T_{env} where ΔT(0) is the initial temperature difference at time 0,

then the Newtonian solution is written as:

$$\Delta T(t) = \Delta T(0)e^{-rt} = \Delta T(0)e^{-t/t_0}.$$

This same solution is more immediately apparent if the initial differential equation is written in terms of ΔT(t), as a single function of time to be found, or "solved for." '

$$\frac{dT(t)}{dt} = \frac{dDT(t)}{dt} = \frac{1}{t_0}\Delta T(t)$$

ELECTROMAGNETIC RADIATION

THERMAL RADIATION

Thermal radiation is electromagnetic radiation generated by the thermal motion of charged particles in matter. All matter with a temperature greater than absolute zero emits thermal radiation. The mechanism is that bodies with a temperature above absolute zero have atoms or molecules with kinetic energies which are changing, and these changes result in charge-acceleration and/or dipole oscillation of the charges that compose the atoms. This motion of charges produces electromagnetic radiation in the usual way. However, the side spectrum of this radiation reflects the wide spectrum of energies and accelerations of the charges in any piece of matter at even a single temperature.

Examples of thermal radiation include the visible light and infrared light emitted by an incandescent light bulb, the infrared radiation emitted by animals and detectable with an infrared camera, and the cosmic microwave background radiation. Thermal radiation is different from thermal convection and thermal conduction—a person near a raging bonfire feels radiant heating from the fire, even if the surrounding air is very cold.

Sunlight is thermal radiation generated by the hot plasma of the Sun. The Earth also emits thermal radiation, but at a much lower intensity and different spectral distribution (infrared rather than visible) because it is cooler. The Earth's absorption of solar radiation, followed by its outgoing thermal radiation are the two most important processes that determine the temperature and climate of the Earth.

If a radiation-emitting object meets the physical characteristics of a black body in thermodynamic equilibrium, the radiation is called blackbody radiation. Planck's law describes the spectrum of blackbody radiation, which depends only on the object's temperature. Wien's displacement law determines the most likely frequency of the emitted radiation, and the Stefan–Boltzmann law gives the radiant intensity.

In engineering, thermal radiation is considered one of the fundamental methods of heat transfer, although a physicist would likely consider energy transfer through thermal radiation a case of one system performing work on another via electromagnetic radiation, and say that heat is a transfer of energy that does not work. The difference is strictly one of nomenclature. Thermal radiation is the emission of electromagnetic waves from all matter that has a temperature greater than absolute zero. It represents a conversion of thermal energy into electromagnetic energy.

Thermal energy results in kinetic energy in the random movements of atoms and molecules in matter. All matter with a temperature by definition is composed of particles which have kinetic energy, and which interact with each other. These atoms and molecules are composed of charged particles, *i.e.,* protons and electrons and kinetic interactions among matter particles result in charge-acceleration and dipole-oscillation. This results in the electrodynamic generation of coupled electric and magnetic fields, resulting in the emission of photons, radiating energy away from the body through its surface boundary. Electromagnetic radiation, or light, does not require the presence of matter to propagate and travels in the vacuum of space infinitely far if unobstructed.

The characteristics of thermal radiation depends on various properties of the surface it is emanating from, including its temperature, its spectral absorptivity and spectral emissive power, as expressed by Kirchhoff's law. The radiation is not monochromatic, *i.e.* it does not consist of just a single frequency, but comprises a continuous dispersion of photon energies, its characteristic spectrum.

If the radiating body and its surface are in thermodynamic equilibrium and the surface has perfect absorptivity at all wavelengths, it is characterized as a black body. A black body is also a perfect emitter. The radiation of such perfect emitters is called black-body radiation. The ratio of any body's emission relative to that of a black body is the body's emissivity, so that a black body has an emissivity of unity. Absorptivity, reflectivity, and emissivity of all bodies are dependent on the wavelength of the radiation. The temperature determines the wavelength distribution of the electromagnetic radiation. For example, fresh snow, which is highly reflective to visible light (reflectivity about 0.90), appears white due to reflecting sunlight with a peak wavelength of about 0.5 micrometers. Its emissivity, however, at a temperature of about –5 °C, peak wavelength of about 12 micrometers, is 0.99.

The distribution of power that a black body emits with varying frequency is described by Planck's law. At any given temperature, there is a frequency f_{max} at which the power emitted is a maximum. Wien's displacement law, and the fact that the frequency of light is inversely proportional to its wavelength in vacuum, mean that the peak frequency f_{max} is proportional to the absolute temperature T of the black body. The photosphere of the Sun, at a temperature of approximately 6000 K, emits radiation principally in the visible portion of the electromagnetic spectrum. Earth's atmosphere is partly transparent to visible light, and the light reaching the surface is absorbed or reflected. Earth's surface emits the absorbed radiation, approximating the behaviour of a black body at 300 K with spectral peak at f_{max}. At these lower frequencies, the atmosphere is largely opaque and radiation from Earth's surface is absorbed or scattered by the atmosphere.

Though some radiation escapes into space, it is absorbed and subsequently re-emitted by atmospheric gases. It is this spectral selectivity of the atmosphere that is responsible for the planetary greenhouse effect, contributing to global warming and climate change in general. The common household incandescent light bulb has a spectrum overlapping the black body spectra of the sun and the earth. A portion of the photons emitted by a tungsten light bulb filament at 3000 K are in the visible spectrum. However, most of the energy is associated with photons of longer wavelengths; these do not help a person see, but still transfer heat to the environment, as can be deduced empirically by observing a household incandescent light bulb. Whenever EM radiation is emitted and then absorbed, heat is transferred. This principle is used in microwave ovens, laser cutting, and RF hair removal.

Unlike conductive and convective forms of heat transfer, thermal radiation can be concentrated in a tiny spot by using reflecting mirrors. Concentrating solar power takes advantage of this fact. In many such systems, mirrors are employed to concentrate sunlight into a smaller area. In lieu of mirrors, Fresnel lenses can also be used to concentrate heat flux. Either method can be used to quickly vaporize water into steam using sunlight. For example, the sunlight reflected from mirrors heats the PS10 Solar Power Plant, and during the day it can heat water to 285 °C (558.15 K) or 545 °F.

Surface Effects

Lighter colours and also whites and metallic substances absorb less illuminating light, and thus heat up less; but otherwise colour makes small difference as regards heat transfer between an object at everyday temperatures and its surroundings, since the dominant emitted wavelengths are nowhere near the visible spectrum, but rather in the far infrared. Emissivities at those wavelengths have little to do with visual emissivities (visible colours); in the far infra-red, most objects have high emissivities. Thus, except in sunlight, the colour of clothing makes little difference as regards warmth; likewise, paint colour of houses makes little difference to warmth except when the painted part is sunlit.

The main exception to this is shiny metal surfaces, which have low emissivities both in the visible wavelengths and in the far infrared. Such surfaces can be used to reduce heat transfer in both directions; an example of this is the multi-layer insulation used to insulate spacecraft.

Low-emissivity windows in houses are a more complicated technology, since they must have low emissivity at thermal wavelengths while remaining transparent to visible light.

Properties

There are four main properties that characterize thermal radiation:

1. Thermal radiation emitted by a body at any temperature consists of a wide range of frequencies. The frequency distribution is given by Planck's law of black-body radiation for an idealized emitter. This is shown in the right-hand diagram.
2. The dominant frequency (or colour) range of the emitted radiation shifts to higher frequencies as the temperature of the emitter increases. For example, a *red hot* object radiates mainly in the long

wavelengths (red and orange) of the visible band. If it is heated further, it also begins to emit discernible amounts of green and blue light, and the spread of frequencies in the entire visible range cause it to appear white to the human eye; it is *white hot*. However, even at a white-hot temperature of 2000 K, 99 per cent of the energy of the radiation is still in the infrared. This is determined by Wien's displacement law. In the diagram the peak value for each curve moves to the left as the temperature increases.

3. The total amount of radiation of all frequencies increases steeply as the temperature rises; it grows as T^4, where T is the absolute temperature of the body. An object at the temperature of a kitchen oven, about twice the room temperature on the absolute temperature scale (600 K vs. 300 K) radiates 16 times as much power per unit area. An object at the temperature of the filament in an incandescent light bulb—roughly 3000 K, or 10 times room temperature—radiates 10,000 times as much energy per unit area. The total radiative intensity of a black body rises as the fourth power of the absolute temperature, as expressed by the Stefan–Boltzmann law. In the plot, the area under each curve grows rapidly as the temperature increases.
4. The rate of electromagnetic radiation emitted at a given frequency is proportional to the amount of absorption that it would experience by the source. Thus, a surface that absorbs more red light thermally radiates more red light. This principle applies to all properties of the wave, including wavelength (colour), direction, polarization, and even coherence, so that it is quite possible to have thermal radiation which is polarized, coherent, and directional, though polarized and coherent forms are fairly rare in nature.

These properties apply if the distances considered are much larger than the wavelengths contributing to the spectrum (most significant from 8-25 micrometres at 300 K). Indeed, thermal radiation here takes only traveling waves into account.

A more sophisticated framework involving electromagnetic theory has to be used for lower distances (near-field thermal radiation).

°C	**Subjective Colour**
480	Faint red glow
580	Dark red
730	Bright red, slightly orange
930	Bright orange
1100	Pale yellowish orange
1300	Yellowish white
> 1400	White (yellowish if seen from a distance through atmosphere)

PRINCIPLE MECHANISMS OF HEAT TRANSFER

INTERCHANGE OF ENERGY

Thermal radiation is one of the principle mechanisms of heat transfer. It entails the emission of a spectrum of electromagnetic radiation due to an object's temperature. Other mechanisms are convection and conduction. The interplay of energy exchange by thermal radiation is characterized by the following equation:

$$\alpha + \rho + \tau = 1$$

Here, α represents the spectral absorption component, ρ spectral reflection component and τ the spectral transmission component. These elements are a function of the wavelength (λ) of the electromagnetic radiation. The spectral absorption is equal to the emissivity ε; this relation is known as Kirchhoff's law of thermal radiation. An object is called a black body if, for all frequencies, the following formula applies:

$$\alpha = \varepsilon = 1$$

In a practical situation and room-temperature setting, humans lose considerable energy due to thermal radiation. However, the energy lost by emitting infrared light is partially regained by absorbing the heat flow due to conduction from surrounding objects, and the remainder resulting from generated heat through metabolism. Human skin has an emissivity of very close to 1.0. Using the formulas below shows a human, having roughly 2 square meter in surface area, and a temperature of about 307 K, continuously radiates approximately 1000 watts.

However, if people are indoors, surrounded by surfaces at 296 K, they receive back about 900 watts from the wall, ceiling, and other surroundings, so the net loss is only about 100 watts. These heat transfer estimates are highly dependent on extrinsic variables, such as wearing clothes, *i.e.* decreasing total thermal circuit conductivity, therefore reducing total output heat flux. Only truly *gray* systems (relative equivalent emissivity/absorptivity and no directional transmissivity dependence in *all* control volume bodies considered) can achieve reasonable steady-state heat flux estimates through the Stefan-Boltzmann law.

Encountering this "ideally calculable" situation is virtually impossible (although common engineering procedures surrender the dependency of these unknown variables and "assume" this to be the case). Optimistically, these "gray" approximations will get you *close* to real solutions, as most divergence from Stefan-Boltzmann solutions is very small (especially in most STP lab controlled environments). If objects appear white (reflective in the visual spectrum), they are not necessarily equally reflective (and thus non-emissive) in the thermal infrared; *e.g.*, most household radiators are painted white despite the fact that they have to be good thermal radiators. Acrylic and urethane based white paints have 93 per cent blackbody radiation efficiency at room temperature (meaning the term "black body" does not always correspond to the visually perceived colour of an object). These materials that do not follow the "black colour = high emissivity/absorptivity" caveat will most likely have functional spectral emissivity/absorptivity dependence.

Calculation of radiative heat transfer between groups of object, including a 'cavity' or 'surroundings' requires solution of a set of simultaneous equations using the Radiosity method. In these calculations, the geometrical configuration of the problem is distilled to a set of numbers called view factors, which give the proportion of radiation leaving any given surface that hits another specific surface. These calculations are important in the fields of solar thermal energy, boiler and furnace design and raytraced computer graphics. A selective surface can be used when energy is being extracted from the sun. For instance, when a green house is made, most of the roof and walls are made out of glass. Glass is transparent in the visible (approximately 0.4 μm $< \lambda <$ 0.8 μm) and near-infrared wavelengths, but opaque to mid- to far-wavelength infrared (approximately $\lambda >$ 3 μm). Therefore glass lets in radiation in the visible range, allowing us to be able to see through it, but doesn't let out radiation that is emitted from objects at or close to room temperature. This traps what we feel as heat. This is known as the greenhouse effect and can be observed by getting into a car that has been sitting in the sun. Selective surfaces can also be used on solar collectors. We can find out how much help a selective surface coating is by looking at the equilibrium temperature of a plate that is being heated through solar radiation. If the plate is receiving a solar irradiation of 1350 W/m^2 (minimum is 1325 W/m^2 on July 4 and maximum is 1418 W/m^2 on January 3) from the sun the temperature of the plate where the radiation leaving is equal to the radiation being received by the plate is 393 K (248 °F).

If the plate has a selective surface with an emissivity of 0.9 and a cut off wavelength of 2.0 μm, the equilibrium temperature is approximately 1250 K (1790 °F). Note that the calculations were made neglecting convective heat transfer and neglecting the solar irradiation absorbed in the clouds/ atmosphere for simplicity, however, the theory is still the same for an actual problem. If we have a surface, such as a glass window, with which we would like to reduce the heat transfer from, a clear reflective film with a low emissivity coating can be placed on the interior of the wall. "Low-emittance (low-E) coatings are microscopically thin, virtually invisible, metal or metallic oxide layers deposited on a window or skylight glazing surface primarily to reduce the U-factor by suppressing radiative heat flow". By adding this coating we are limiting the amount of radiation that leaves the window thus increasing the amount of heat that is retained inside the window.

Radiative Heat Transfer

The radiative heat transfer from one surface to another is equal to the radiation entering the first surface from the other, minus the radiation leaving the first surface.

- *For a black body:*

$$\dot{Q}_{1\to2} = A_1 E_{b1} F_{1\to2} - A_2 E_{b2} F_{2\to1}$$

Using the reciprocity rule, $A_1 F_{1\to2} = A_2 F_{2\to1}$, *this simplifies to:*

$$\dot{Q}_{1\to2} = \sigma A_1 F_{1\to2} \left(T_1^4 - T_2^4\right)$$

where σ is the Stefan–Boltzmann constant.

- *For a grey body with only two surfaces the heat transfer is equal to:*

$$\dot{Q} = \frac{\sigma\left(T_1^4 - T_2^4\right)}{\frac{1-\varepsilon_1}{A_1\varepsilon_1} + \frac{1}{A_1 F_{12}} + \frac{1-\varepsilon_2}{A_2\varepsilon_2}}$$

However, this value can easily change for different circumstances and different equations should be used on a case per case basis.

Radiative Power

Thermal radiation power of a black body per unit of solid angle and per unit frequency V is given by Planck's law as:

$$u(v,T) = \frac{2hv^3}{c^2} \cdot \frac{1}{e^{hv/k_BT} - 1}$$

or,

$$u(\lambda,T) = \frac{\beta}{\lambda^5} \cdot \frac{1}{e^{hc/k_BT\lambda} - 1}$$

where β is a constant.

This formula mathematically follows from calculation of spectral distribution of energy in quantized electromagnetic field which is in complete thermal equilibrium with the radiating object. The equation is derived as an infinite sum over all possible frequencies. The energy, E = hv, of each photon is multiplied by the number of states available at that frequency, and the probability that each of those states will be occupied.

Integrating the above equation over V the power output given by the Stefan–Boltzmann law is obtained, as:

$$P = \sigma \cdot A \cdot T^4$$

where the constant of proportionality σ is the Stefan–Boltzmann constant and A is the radiating surface area.

Further, the wavelength λ, for which the emission intensity is highest, is given by Wien's Law as:

$$\lambda_{max} = \frac{b}{T}$$

For surfaces which are not black bodies, one has to consider the (generally frequency dependent) emissivity factor ε(v). This factor has to be multiplied with the radiation spectrum formula before integration. If it is taken as a constant, the resulting formula for the power output can be written in a way that contains ε as a factor:

$$P = \varepsilon \cdot \sigma \cdot A \cdot T^4$$

This type of theoretical model, with frequency-independent emissivity lower than that of a perfect black body, is often known as a *gray body*. For frequency-dependent emissivity, the solution for the integrated power depends on the functional form of the dependence, though in general there is no simple expression for it.

Practically speaking, if the emissivity of the body is roughly constant around the peak emission wavelength, the gray body model tends to work fairly well since the weight of the curve around the peak emission tends to dominate the integral.

Constants

Definitions of constants used in the above equations:

h	Planck's constant	6.626 0693(11) × 10^{-34} J·s = 4.135 667 43(35) × 10^{-15} eV·s
b	Wien's displacement constant	2.897 7685(51) × 10^{-3} m·K
k_B	Boltzmann constant	1.380 6505(24) × 10^{-23} J·K^{-1} = 8.617 343(15) × 10^{-5} eV·K^{-1}
s	Stefan–Boltzmann constant	5.670 373(21) × 10^{-8} W·m^{-2}·K^{-4}
C	Speed of light	299,792.458 m·s^{-1}

Variables

Definitions of variables, with example values:

T	Absolute temperature	For units used above, must be in kelvin (*e.g.* Average surface temperature on Earth = 288 K)
A	Surface area	$A_{cuboid} = 2ab + 2bc + 2ac$; $A_{cylinder} = 2\pi r(h + r)$; $A_{sphere} = 4\pi r$

TRANSFERRING MATTER OF ADVECTION

By transferring matter, energy—including thermal energy—is moved by the physical transfer of a hot or cold object from one place to another. This can be as simple as placing hot water in a bottle and heating a bed, or the movement of an iceberg in changing ocean currents. A practical example is thermal hydraulics. This can be described by the formula,

$$Q = v \cdot r \cdot c_p \cdot \Delta T$$

Convection vs. Conduction

In a body of fluid that is heated from underneath its container, conduction and convection can be considered to compete for dominance. If heat conduction is too great, fluid moving down by convection is heated by conduction so fast that its downward movement will be stopped due to its buoyancy, while fluid moving up by convection is cooled by conduction so fast that its driving buoyancy will diminish. On the other hand, if heat conduction is very low, a large temperature gradient may be formed and convection might be very strong.

The Rayleigh number (Ra) is a measure determining the result of this competition.

$$Ra = \frac{gDrL^3}{\mu\alpha} = \frac{g\beta\Delta TL^3}{\nu\alpha}$$

where,

- *g* is acceleration due to gravity
- ρ is the density with Δρ being the density difference between the lower and upper ends
- μ is the dynamic viscosity
- α is the Thermal diffusivity
- β is the volume thermal expansivity (sometimes denoted *á* elsewhere)
- *T* is the temperature and
- ν is the kinematic viscosity.

The Rayleigh number can be understood as the ratio between the rate of heat transfer by convection to the rate of heat transfer by conduction; or, equivalently, the ratio between the corresponding timescales (*i.e.* conduction timescale divided by convection timescale), up to a numerical factor. This can be seen as follows, where all calculations are up to numerical factors depending on the geometry of the system.

The buoyancy force driving the convection is roughly $g\Delta\rho L^3$, so the corresponding pressure is roughly $g\Delta\rho L$. In steady state, this is canceled by the shear stress due to viscosity, and therefore roughly equals $\mu V/L = \mu/T_{conv}$, where *V* is the typical fluid velocity due to convection and T_{conv}

the order of its timescale. The conduction timescale, on the other hand, is of the order of $T_{cond} = L^2/\alpha$.

Convection occurs when the Rayleigh number is above 1,000–2,000. For example, the Earth's mantle, exhibiting non-stable convection, has Rayleigh number of the order of 1,000, and T_{conv} as calculated above is around 100 million years.

Phase Changes

Transfer of heat through a phase transition in the medium—such as water-to-ice, water-to-steam, steam-to-water, or ice-to-water—involves significant energy and is exploited in many ways: steam engines, refrigerators, etc. For example, the Mason equation is an approximate analytical expression for the growth of a water droplet based on the effects of heat transport on evaporation and condensation.

Boiling

Heat transfer in boiling fluids is complex, but of considerable technical importance. It is characterized by an *S*-shaped curve relating heat flux to surface temperature difference.

At low driving temperatures, no boiling occurs and the heat transfer rate is controlled by the usual single-phase mechanisms. As the surface temperature is increased, local boiling occurs and vapour bubbles nucleate, grow into the surrounding cooler fluid, and collapse. This is *sub-cooled nucleate boiling,* and is a very efficient heat transfer mechanism.

At high bubble generation rates, the bubbles begin to interfere and the heat flux no longer increases rapidly with surface temperature (this is the departure from nucleate boiling, or DNB). At higher temperatures still, a maximum in the heat flux is reached (the critical heat flux, or CHF). The regime of falling heat transfer that follows is not easy to study, but is believed to be characterized by alternate periods of nucleate and film boiling. Nucleate boiling slows the heat transfer due to gas bubbles on the heater's surface; as mentioned, gas-phase thermal conductivity is much lower than liquid-phase thermal conductivity, so the outcome is a kind of "gas thermal barrier".

At higher temperatures still, the hydrodynamically-quieter regime of film boiling is reached. Heat fluxes across the stable vapour layers are low, but rise slowly with temperature. Any contact between fluid and the surface that may be seen probably leads to the extremely rapid nucleation of a fresh vapour layer ("spontaneous nucleation").

CONDENSATION HEAT TRANSFER

Condensation occurs when a vapour is cooled and changes its phase to a liquid. Condensation heat transfer, like boiling, is of great significance in industry. During condensation, the latent heat of vaporization must be released. The amount of the heat is the same as that absorbed during vaporization at the same fluid pressure.

There are several types of condensation:

- Homogeneous condensation, as during a formation of fog.
- Condensation in direct contact with subcooled liquid.
- Condensation on direct contact with a cooling wall of a heat exchanger: This is the most common mode used in industry:
  - Filmwise condensation is when a liquid film is formed on the subcooled surface, and usually occurs when the liquid wets the surface.
  - Dropwise condensation is when liquid drops are formed on the subcooled surface, and usually occurs when the liquid does not wet the surface.

Dropwise condensation is difficult to sustain reliably; therefore, industrial equipment is normally designed to operate in filmwise condensation mode.

Modeling Approaches

Complex heat transfer phenomena can be modeled in different ways.

Heat Equation

The heat equation is an important partial differential equation that describes the distribution of heat (or variation in temperature) in a given region over time. In some cases, exact solutions of the equation are available; in other cases the equation must be solved numerically using computational methods. For example, simplified climate models may use Newtonian cooling, instead of a full (and computationally expensive) radiation code, to maintain atmospheric temperatures.

LUMPED SYSTEM ANALYSIS

System analysis by the lumped capacitance model is a common approximation in transient conduction that may be used whenever heat conduction within an object is much faster than heat conduction across the boundary of the object. This is a method of approximation that reduces one aspect of the transient conduction system—that within the object—to an equivalent steady state system. That is, the method assumes that the temperature within the object is completely uniform, although its value may be changing in time.

In this method, the ratio of the conductive heat resistance within the object to the convective heat transfer resistance across the object's boundary, known as the *Biot number*, is calculated. For small Biot numbers, the approximation of *spatially uniform temperature within the object* can be used: it can be presumed that heat transferred into the object has time to uniformly distribute itself, due to the lower resistance to doing so, as compared with the resistance to heat entering the object. Lumped system analysis often reduces the complexity of the equations to one first-order linear differential equation, in which case heating and cooling are described by a simple exponential solution, often referred to as Newton's law of cooling.

APPLICATIONS AND TECHNIQUES

Heat transfer has broad application to the functioning of numerous devices and systems. Heat-transfer principles may be used to preserve, increase, or decrease temperature in a wide variety of circumstances.

Insulation and Radiant Barriers

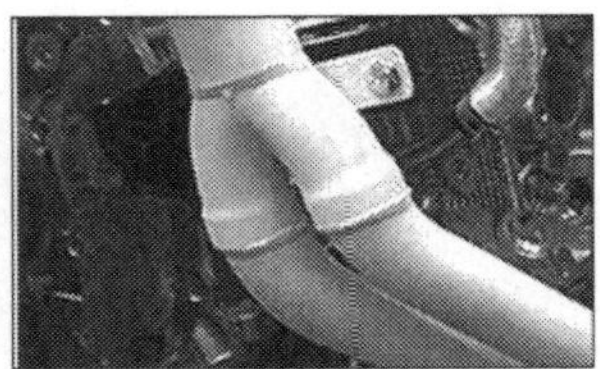

Fig. Car Exhausts usually Require some form of Heat Barrier, Especially high Performance Exhausts where a Ceramic Coating is often Applied.

Fig. Heat Exposure as part of a Fire test for Firestop Products.

Thermal insulators are materials specifically designed to reduce the flow of heat by limiting conduction, convection, or both. Radiant barriers are materials that reflect radiation, and therefore reduce the flow of heat from radiation sources. Good insulators are not necessarily good radiant barriers, and vice versa. Metal, for instance, is an excellent reflector and a poor insulator.

The effectiveness of an insulator is indicated by its R-value, or resistance value. The R-value of a material is the inverse of the conduction coefficient (*k*) multiplied by the thickness (*d*) of the insulator. In most of the world, R-values are measured in SI units: square-meter kelvins per watt ($m^2 \cdot K/W$). In the United States, R-values are customarily given in units of British thermal units per hour per square-foot degrees Fahrenheit ($Btu/h \cdot ft^2 \cdot {}^\circ F$).

$$R = \frac{d}{k}$$

$$C = \frac{Q}{m\Delta T}$$

Rigid fiberglass, a common insulation material, has an R-value of four per inch, while poured concrete, a poor insulator, has an R-value of 0.08 per inch. The tog is a measure of thermal resistance, commonly used in the textile industry, and often seen quoted on, for example, duvets and carpet underlay.

The effectiveness of a radiant barrier is indicated by its reflectivity, which is the fraction of radiation reflected. A material with a high reflectivity (at a given wavelength) has a low emissivity (at that same wavelength), and vice versa. At any specific wavelength, reflectivity = 1 - emissivity. An ideal radiant barrier would have a reflectivity of 1, and would therefore reflect 100 per cent of incoming radiation.

Vacuum flasks, or Dewars, are silvered to approach this ideal. In the vacuum of space, satellites use multi-layer insulation, which consists of many layers of aluminized (shiny) Mylar to greatly reduce radiation heat transfer and control satellite temperature.

Critical Insulation Thickness

Low thermal conductivity (k) materials reduce heat fluxes. The smaller the k value, the larger the corresponding thermal resistance (R) value. Thermal conductivity is measured in watts-per-meter per kelvin ($W \cdot m^{-1} \cdot K^{-1}$), represented as k. As the thickness of insulating material increases, the thermal resistance—or R-value—also increases.

However, adding layers of insulation has the potential of increasing the surface area, and hence the thermal convection area.

For example, as thicker insulation is added to a cylindrical pipe, the outer radius of the pipe-and-insulation system increases, and therefore surface area increases. The point where the added resistance of increasing insulation thickness becomes overshadowed by the effect of increased surface area is called the critical insulation thickness. In simple cylindrical pipes, this is calculated as a radius:

$$R_{critical} = \frac{k}{h}$$

EFFICIENT HEAT TRANSFER

Heat Exchangers

A heat exchanger is a tool built for efficient heat transfer from one fluid to another, whether the fluids are separated by a solid wall so that they never mix, or the fluids are in direct contact. Heat exchangers are widely used in refrigeration, air conditioning, space heating, power generation, and chemical processing. One common example of a heat exchanger is a car's radiator, in which the hot coolant fluid is cooled by the flow of air over the radiator's surface.

Common types of heat exchanger flows include parallel flow, counter flow, and cross flow. In parallel flow, both fluids move in the same direction while transferring heat; in counter flow, the fluids move in opposite directions; and in cross flow, the fluids move at right angles to each other. Common constructions for heat exchanger include shell and tube, double pipe, extruded finned pipe, spiral fin pipe, u-tube, and stacked plate.

When engineers calculate the theoretical heat transfer in a heat exchanger, they must contend with the fact that the driving temperature difference between the two fluids varies with position. To account for this in simple systems, the log mean temperature difference (LMTD) is often used as an "average" temperature. In more complex systems, direct knowledge of the LMTD is not available, and the number of transfer units (NTU) method can be used instead.

Heat Dissipation

A heat sink is a component that transfers heat generated within a solid material to a fluid medium, such as air or a liquid. Examples of heat sinks are the heat exchangers used in refrigeration and air conditioning systems, and the radiator in a car (which is also a heat exchanger). Heat sinks also help to cool electronic and optoelectronic devices such as CPUs, higher-power lasers, and light-emitting diodes (LEDs). A heat sink uses its extended surfaces to increase the surface area in contact with the cooling fluid.

Buildings

In cold climates, houses with their heating systems form dissipative systems, often resulting in a loss of energy (known colloquially as "Heat Bleed") that makes home interiors uncomfortably cool or cold.

For the comfort of the inhabitants, the interiors must be maintained out of thermal equilibrium with the external surroundings. In effect, these domestic residences are islands of warmth in a sea of cold, and the thermal gradient between the inside and outside is often quite steep. This can lead to problems such as condensation and uncomfortable air currents, which—if left unaddressed—can cause cosmetic or structural damage to the property.

Such issues can be prevented through the execution of an energy audit, and the implementation of recommended corrective procedures (such as the installation of adequate insulation, the air sealing of structural leaks, and the addition of energy-efficient windows and doors.

Thermal transmittance is the rate of transfer of heat through a structure divided by the difference in temperature across the structure. It is expressed in watts per square meter per kelvin, or W/m^2K. Well-insulated parts of a building have a low thermal transmittance, whereas poorly-insulated parts of a building have a high thermal transmittance.

A thermostat is a device capable of starting the heating system when the house's interior falls below a set temperature, and of stopping that same system when another (higher) set temperature has been achieved. Thus, the thermostat controls the flow of energy into the house, that energy eventually being dissipated to the exterior.

Thermal Energy Storage

Thermal energy storage refers to technologies that store energy in a thermal reservoir for later use. They can be employed to balance energy demand between daytime and nighttime. The thermal reservoir may be maintained at a temperature above (hotter) or below (colder) than that of the ambient environment. Applications include later use in space heating, domestic or process hot water, or to generate electricity. Most practical active solar heating systems have storage for a few hours to a day's worth of heat collected.

Evaporative Cooling

Evaporative cooling is a physical phenomenon in which evaporation of a liquid, typically into surrounding air, cools an object or a liquid in contact with it. Latent heat describes the amount of heat that is needed to evaporate the liquid; this heat comes from the liquid itself and the surrounding gas and surfaces. The greater the difference between the two temperatures, the greater the evaporative cooling effect. When the temperatures are the same, no net evaporation of water in air occurs; thus, there is no cooling effect. A simple example of natural evaporative cooling is perspiration, or sweat, which the body secretes in order to cool itself. An evaporative cooler is a device that cools air through the simple evaporation of water.

Radiative Cooling

Radiative cooling is the process by which a body loses heat by radiation. It is an important effect in the Earth's atmosphere. In the case of the Earth-atmosphere system, it refers to the process by which long-wave (infrared) radiation is emitted to balance the absorption of short-wave (visible) energy from the Sun. Convective transport of heat and evaporative transport of latent heat both remove heat from the surface and redistribute it in the atmosphere, making it available for radiative transport at higher altitudes.

Laser Cooling

Laser cooling refers to techniques in which atomic and molecular samples are cooled through the interaction with one or more laser light fields. The most common method of laser cooling is Doppler cooling. In Doppler cooling, the frequency of the laser light is tuned slightly below an electronic transition in the atom. Thus, the atoms would absorb more photons if they moved towards the light source, due to the Doppler effect.

If an excited atom then emits a photon spontaneously, it will be accelerated. The result of the absorption and emission process is to reduce the speed of the atom. Eventually the mean velocity, and therefore the kinetic energy of the atoms, will be reduced. Since the temperature of an ensemble of atoms is a measure of the random internal kinetic energy, this is equivalent to cooling the atoms.

Sympathetic cooling is a process in which particles of one type cool particles of another type. Typically, atomic ions that can be directly laser-cooled are used to cool nearby ions or atoms. This technique allows cooling of ions and atoms that cannot be laser cooled directly.

Magnetic Cooling

Magnetic evaporative cooling is a technique for lowering the temperature of a group of atoms. The process confines atoms using a magnetic field. Over time, individual atoms will become much more energetic than the others due to random collisions, and will escape—removing energy from the system and reducing the temperature of the remaining group. This process is similar to the familiar process by which standing water becomes water vapour.

4

Physical Concept of Entropy

Entropy is a thermodynamic property that is a measure of the energy not available for useful work in a thermodynamic process, such as in energy conversion devices, engines, or machines. Such devices can only be driven by convertible energy, and have a theoretical maximum efficiency when converting energy to work. During this work entropy accumulates in the system, but has to be removed by dissipation in the form of waste heat.

PHYSICAL SIGNIFICANCE OF ENTROPY

ENTROPY AS A MEASURE OF THE DISORDER OF THE SYSTEM

We know that all the spontaneous process in which heat is transferred through a finite temperature. Spontaneous process is also known as irreversible process because these processes take place at very fast rate. All the natural processes are spontaneous processes. All these processes are accompanied by increase in the disorder of the system.

For example: When we heat a metallic rod from one end the initially all the molecules are present in some order. But when, heat transfer take place from hot end to cold end disorder of the molecules take place. This is due to attainment of the equilibrium which will make uniform distribution of the energy in the system. Hence the state of the order of the molecules gets disturbed and gives rise to greater disorder. Hence spontaneous processes are accompanied by increase in entropy as well as increase in the disorder of the system.

Entropy as a measure of Probability

From above we concluded that spontaneous processes are accompanied by increase in entropy as well as increase in the disorder of the system. A little consideration will show that when a spontaneous process takes place it moves from less probable state to a more probable state. Hence, there is a close relationship between entropy S and the thermodynamic probability W of the state of the system. This relationship is expressed by Boltzmann as:

$S = k \ln W + \text{constant}$

Where k denotes the Boltzmann constant which is equal to gas constant per single molecule i.e. R/NA.

As constant is equal to zero, therefore:

$S = k \ln W$

This is called the Boltzmann entropy equation.

CONCEPT OF ENTROPY

Entropy is a chemical concept that is very difficult to explain, because a one-sentence definition will not lead to a comprehensive statement.

The concept of entropy is fundamental to understanding the second law of thermodynamics. Entropy (or more specifically, increase in entropy) is defined as heat (in calories or Btu's) absorbed by a system, divided by the absolute temperature of the system at the time the heat is absorbed. Absolute temperature is the number of degrees above "absolute zero", the coldest temperature that can exist.

The total entropy in a system is represented by the symbol ΔS. The symbol S is used to represent a given change in the entropy content of a system. If the symbol q is used to represent the amount of heat absorbed by a system, the equation for the resulting entropy increase is:

$$\Delta S = q/T$$

Where T is the absolute temperature. When heat is absorbed, the entropy of a system increases; when heat flows out of a system, its entropy decreases.

The "surroundings" of a system is everything outside of the system that can interact with it; surroundings can usually be defined as the space that surrounds a system. When heat is evolved by a system, that same heat is absorbed by its surroundings.

When heat is absorbed by a system, that same heat must necessarily come from its surroundings. Therefore any entropy increase in a system due to heat flow must be accompanied by an entropy decrease in the surroundings, and vice versa. When heat flows spontaneously from a hotter region to a cooler region, the entropy decrease in the hotter region will always be less than the entropy increase in the cooler region, because the greater the absolute temperature, the smaller the entropy change for any particular heat flow.

As an example, consider the entropy change when a large rock at 500 degrees absolute is dropped into water at 650 degrees absolute. (We are using an absolute temperature scale based on Fahrenheit degrees; on this scale, water freezes at 492 degrees.) For each Btu of heat that flows into the rock at these temperatures the entropy increase in the rock is 1/500 = 0.0020 and the entropy decrease of the water is 1/650 = 0.0015. The difference between these values is 0.0020 - 0.0015 = 0.0005. This represents the *over all* entropy increase of the system (rock) and its surroundings (water).

Of course the rock will warm up to, and the water cool to, a temperature intermediate between their original temperatures, thus considerably complicating the calculation of total entropy change after equilibrium is achieved. Nevertheless, for every Btu of heat transferred from water to rock there will always be an increase of over-all net entropy.

As was mentioned before, a spontaneous change is an irreversible change. Therefore an increase in the overall net entropy can be used as a measure of the irreversibility of spontaneous heat flow.

Irreversible changes in a system can, and often do, take place even though there may be no interaction, and negligible heat flow, between system and surroundings. In cases like these the entropy "content" of the system is greater after the change than before. Even when heat flow does not occur between system and surroundings, spontaneous changes inside an isolated system are always accompanied by an increase in the system's entropy, and this calculated entropy increase can be used as a measure of irreversibility. The following paragraphs will explain how this entropy increase can, at least in some cases, be calculated.

It is an axiom of thermodynamics that entropy, like temperature, pressure, density, etc., is a property of a system and depends only on the existing condition of the system. Regardless of the procedures followed in achieving a given condition, the entropy content for that condition is always the same. In other words, for any given set of values for pressure, temperature, density, composition, etc., there can be only one value for the entropy content. It is essential to remember this: When a system that has undergone an irreversible change is restored to its original condition (same temperature, pressure, volume, etc.) its entropy content will likewise be the same as it was before.

In cases where an isolated system undergoes an entropy increase as the result of a spontaneous change inside the system, we can calculate that entropy increase by postulating a procedure whereby the system's entropy increase is transferred to the surroundings in a manner such that there is no further increase in *net* entropy and the system is restored to its original condition. The entropy increase of the surroundings can then be readily calculated by equation:

$$\Delta S = q/T,$$

where,

q = heat absorbed by the surroundings,

T = absolute temperature of the surroundings.

It bears repeating that when the system is restored to its original condition, its entropy content will be the same as it was before its irreversible change. Therefore the amount of entropy absorbed by the surroundings during restoration must necessarily be the same as the entropy increase accompanying the system's original irreversible change, providing that there is no further increase in *net* entropy during restoration. This postulated restoration procedure and the postulated properties of the surroundings are for the purpose of calculation only.

Since we are not dealing with the surroundings as such, they can be postulated in whatever form necessary to simplify the calculations; it is neither necessary nor desirable that the surroundings correspond to any condition that could actually exist. Therefore, we will postulate a theoretical restoration procedure that takes place with no further increase in *net* entropy, even though such a procedure can not actually be obtained experimentally.

The restoration process, if it were to take place in actuality, would have to be accompanied by at least a small amount of irreversibility, and hence an additional increase in the entropy of the surroundings beyond the entropy increase from the system's original irreversible change. This is because heat will not flow without a temperature differential, friction cannot be entirely eliminated, etc. Therefore the restoring process, if it is to take place with no further increase in over-all net entropy, must be postulated to take place with no irreversibility.

If such a process could be actually realised, it would be characterized by a continuous state of equilibrium (*i.e.* no pressure or temperature differentials) and would occur at a rate so slow as to require infinite time. Processes like these are called "reversible" processes. Remember, reversible processes are postulated to simplify the calculation of the entropy change in a system; it is *not* necessary that they be capable of being achieved experimentally.

It should not be assumed that equation requires that q, the heat absorbed, must necessarily be absorbed reversibly. The concept of reversibility is merely a means to an end: the calculation of entropy change accompanying an irreversible process.

The following example will illustrate the calculation of a reversible restoring process and at the same time develop the equation which is the basis for the thermodynamical relationship between probability and the second law. We will postulate a system consisting of an "ideal" gas contained in a tank connected to a second tank that has been completely evacuated, with the valve between the two tanks closed. The temperature of the system and its surroundings is postulated to be the same.

An ideal gas is one whose molecules are infinitely small and have no attractive or repulsive forces on each other. (Under ordinary conditions hydrogen and helium closely approximate the properties of an ideal gas.) An ideal gas is chosen in order to develop the basic relationship without introducing complicating correction factors to account for the size of the molecules and the forces they exert on each other.

When the valve is opened the gas expands irreversibly from V_1 (its original volume) to V_2 (the volume of both tanks). There is no work of compression by or upon the surroundings. Because the gas is ideal there is no temperature change, and hence no heat flow takes place. After expanding irreversibly from V_1 to V_2, the gas is restored to its original condition by reversibly compressing it back to V_1.

This compression requires work (force applied through a distance) which in turn generates heat in the gas, heat that is absorbed by the surroundings so that there is no increase in the gas temperature. In our mathematical model of this reversible restoring process, the surroundings are postulated to be so large that they also do not undergo any temperature increase. The temperature T remains unchanged during the entire irreversible expansion and subsequent reversible restoration process.

The work of compressing the gas during restoration is equal to the pressure of the gas times the volume change due to compression. Because the pressure increases during compression, the work of compression must be determined by the calculus integral:

$$\text{Compression Work} = \int P dV$$

where:

P = pressure

V = volume

dV = the small change in volume taking place at the corresponding pressure P.

The integral sign indicates the summation of all the individual values of PdV.

The equation relating temperature, pressure, and volume of an ideal gas is:

$$PV = RT$$

where:

P = pressure

V = volume

T = absolute temperature

R = a constant which depends only on the amount of gas present.

In the case of a reversible, isothermal compression of an ideal gas we may substitute P from equation into the equation for compression work. When this is done, *we have*:

$$\text{Compression Work} = RT \int dV/V$$

Although it is not necessary that our postulated reversible restoration process be capable of being carried out in a practical sense, it is nevertheless sometimes helpful to be able to visualize the process. To this end, the reader may consider the restoring compression process being brought about by a piston fitted into the end of the second tank. On compression from V_2 to V_1, the piston moves down the length of the second tank, and with no mechanical friction forces all the gas contained therein back into the first tank V_1.

Since the work of compression is equal to q, the heat absorbed by the surroundings, q may be substituted in equation to give:

$$q = RT \int dV/V$$

From equation the entropy gained by the surroundings during restoration from V_2 to V_1 is:

$$\Delta S = q/T$$

Substituting from equation:

$$\Delta S = RdV/V$$

Upon integrating (a calculus procedure for summing up the individual values of dV/V) we have:

$$\Delta S = R\ln(V_2/V_1)$$

Where $\ln(V_2/V_1)$ is the natural logarithm of the ratio of expanded volume to the initial volume, and ΔS is equal to the entropy increase in the surroundings upon restoration compression from V_2 to V_1. As we have seen, ΔS is also equal to the entropy increase of the gas caused by its original expansion from V_1 to V_2. This is because V_1 is the same volume both before expansion and after restoration compression, and therefore has the same entropy content. Therefore the entropy transferred to the surroundings during restoration is equal to that gained by the system in expanding from V_1 to V_2.

PROPERTIES OF ENTROPY

Entropy, as we have defined it, has some dependence on the resolution dE to which the energy of macrostates is measured. Recall that W(E) is the number of accessible microstates with energy in the range E to E + dE. Suppose that we choose a new resolution d*E and define a new density of states W* E which is the number of states with energy in the range E to E + d*E. It can easily be seen that,

$$\Omega^*(E) = \frac{\delta^* E}{\delta E}\Omega(E).$$

It follows that the new entropy S= k ln W* is related to the previous entropy S* = k ln W via:*

$$S^* = S + k\ln\frac{\delta^* E}{\delta E}.$$

Now, our usual estimate that W ~ E^f gives $S \sim k^f$, where f is the number of degrees of freedom. It follows that even if d*E were to differ from dE by of order f (*i.e.*, twenty four orders of magnitude), which is virtually inconceivable, the second term on the right-hand side of the above equation is still only of order $k \ln f$, which is utterly negligible compared to kf. It follows that

$$S^* = S$$

to an excellent approximation, so our definition of entropy is completely insensitive to the resolution to which we measure energy (or any other macroscopic parameter).

Note that, like the temperature, the entropy of a macrostate is only well-defined if the macrostate is in equilibrium. The crucial point is that it only makes sense to talk about the number of accessible states if the systems in the ensemble are given sufficient time to thoroughly explore all of the possible microstates consistent with the known macroscopic constraints.

In other words, we can only be sure that a given microstate is inaccessible when the systems in the ensemble have had ample opportunity to move into it, and yet have not done so. Note that for an equilibrium state, the entropy is just as well-defined as more familiar quantities such as the temperature and the mean pressure.

Consider, again, two systems *A* and *A*¢which are in thermal contact but can do no work on one another. Let *E* and *E*¢ be the energies of the two systems, and W¢(*E*) and W¢(*E*¢) the respective densities of states. Furthermore, let $E^{(0)}$ be the conserved energy of the system as a whole and the $W^{(0)}$ corresponding density of states. We have from equation that,

$$\Omega^{(0)}(E) = \Omega(E)\,\Omega'(E'),$$

where *E*¢ = $E^{(0)} - E$. In other words, the number of states accessible to the whole system is the product of the numbers of states accessible to each subsystem, since every microstate of *A* can be combined with every microstate of *A*¢to form a distinct microstate of the whole system that in equilibrium the mean energy of *A* takes the value $\bar{E} = \tilde{E}$ for which $W^{(0)}(E)$ is maximum, and the temperatures of *A* and *A*¢ are equal. The distribution of *E* around the mean value is of order D*$E = \tilde{E}/\sqrt{f}$, where *f* is the number of degrees of freedom. It follows that the total number of accessible microstates is approximately the number of states which lie within D**E* of $\tilde{E}$. Thus,

$$\Omega^{(0)}_{tot} \simeq \frac{\Omega^{(0)}(\tilde{E})}{\delta E}\Delta * E.$$

The entropy of the whole system is given by,

$$S^{(0)} = k\ln\Omega^{(0)}_{tot} = k\ln\Omega^{(0)}(\tilde{E}) + k\ln\frac{\Delta * E}{\delta E}.$$

According to our usual estimate, W ~ E^f, the first term on the right-hand side is of order *kf* whereas the second term is of order *k* ln ($\tilde{E}/\sqrt{f}$ d*E*). Any reasonable choice for the energy subdivision d*E* should be greater than $\tilde{E}/f$, otherwise there would be less than one microstate per subdivision. It follows that the second term is less than or of order *k* ln *f*, which is utterly negligible compared to *kf*. Thus,

$$S^{(0)} = k\ln\Omega^{(0)}(\tilde{E}) = k\ln[\Omega(\tilde{E})\Omega(\tilde{E}')] = k\ln\Omega(\tilde{E}) + k\ln\Omega'(\tilde{E}')$$

to an excellent approximation, giving

$$S^{(0)} = S(\tilde{E}) + S'(\tilde{E}').$$

It can be seen that the probability distribution for $W^{(0)}(E)$is so strongly peaked around its maximum value that, for the purpose of calculating the entropy, the total number of states is equal to the maximum number of states [*i.e.*, $\Omega_{tot}^{(0)} \sim \Omega^{(0)}(\tilde{E})$]. One consequence of this is that the entropy has the simple additive property equation. Thus, the total entropy of two thermally interacting systems in equilibrium is the sum of the entropies of each system in isolation.

Uses of Entropy

We have defined a new function called entropy, denoted S, which parameterizes the amount of disorder in a macroscopic system. The entropy of an equilibrium macrostate is related to the number of accessible microstates W via

$$S = k \ln \Omega.$$

On a macroscopic level, the increase in entropy due to a quasi-static change in which an infinitesimal amount of heat $đQ$ is absorbed by the system is given by

$$dS = \frac{đQ}{T},$$

where T is the absolute temperature of the system. The second law of thermodynamics states that the entropy of an isolated system can never spontaneously decrease. Let us now briefly examine some consequences of these results.

Consider two bodies, A and $A¢$, which are in thermal contact but can do no work on one another. We know what is supposed to happen here. Heat flows from the hotter to the colder of the two bodies until their temperatures are the same. Consider a quasi-static exchange of heat between the two bodies. According to the first law of thermodynamics, if an infinitesimal amount of heat $đQ$ is absorbed by A then infinitesimal heat $đQ¢ = -đQ$ is absorbed by $A¢$. The increase in the entropy of system A is $dS = đQ/T$ and the corresponding increase in the entropy of $A¢$ is $dS = đQ/T¢$. Here, T and $T¢$ are the temperatures of the two systems, respectively. Note that $đQ$ is assumed to the sufficiently small that the heat transfer does not substantially modify the temperatures of either system. The change in entropy of the whole system is

$$dS^{(0)} = dS + dS' = \left(\frac{1}{T} - \frac{1}{T'}\right) đQ,$$

This change must be positive or zero, according to the second law of thermodynamics, so $dS^{(0)}$ 3 0. It follows that $\bar{d}Q$ is positive (*i.e.*, heat flows from $A¢$ to A) when $T¢ > T$, and vice versa. The spontaneous flow of heat only ceases when $T = T¢$. Thus, the direction of spontaneous heat flow is a consequence of the second law of thermodynamics.

Note that the spontaneous flow of heat between bodies at different temperatures is always an irreversible process which increases the entropy, or disorder, of the Universe.

Consider, now, the slightly more complicated situation in which the two systems can exchange heat and also do work on one another via a movable partition. Suppose that the total volume is invariant, so that,

$$V^{(0)} = V + V¢ \text{ constant,}$$

where V and $V¢$ are the volumes of A and $A¢$, respectively. Consider a quasi-static change in which system A absorbs an infinitesimal amount of heat $\bar{d}Q$ and its volume simul-taneously increases by an infinitesimal amount dV. The infinitesimal amount of work done by system A is $\bar{d}W = \bar{p}dV$, where $\bar{p}$ is the mean pressure of A. According to the first law of thermodynamics,

$$\bar{d}Q = dE + \bar{d}W = dE + \bar{p}dV,$$

where dE is the change in the internal energy of A. Since $dS = \bar{d}Q/T$, the increase in entropy of system A is written,

$$dS = \frac{dE + \bar{p}dV}{T}.$$

Likewise, the increase in entropy of system $A¢$ is given by,

$$dS' = \frac{dE' + \bar{p}'dV'}{T'}.$$

According to equation

$$\frac{1}{T} = \left(\frac{\partial S}{\partial E}\right)_V,$$

$$\frac{\bar{p}}{T} = \left(\frac{\partial S}{\partial V}\right)_E,$$

where the subscripts are to remind us what is held constant in the partial derivatives. We can write a similar pair of equations for the system $A¢$.

The overall system is assumed to be isolated, so conservation of energy gives $dE + dE¢= 0$. Furthermore, equation implies that $dV + dV¢= 0$. It follows that the total change in entropy is given by,

$$dS^{(0)} = dS + dS' = \left(\frac{1}{T} - \frac{1}{T'}\right)dE + \left(\frac{\bar{p}}{T} - \frac{\bar{p}'}{T'}\right)dV.$$

The equilibrium state is the most probable state. According to statistical mechanics, this is equivalent to the state with the largest number of accessible microstates. Finally, equation implies that this is the maximum entropy state. The system can never spontaneously leave a maximum entropy state, since this would imply a spontaneous reduction in entropy, which is forbidden by the second law of thermodynamics. A maximum or minimum entropy state must satisfy $dS^{(0)} = 0$ for arbitrary small variations of the energy and external parameters.

It follows from equation that,

$$T = T'$$

$$\bar{p} = \bar{p}',$$

for such a state. This corresponds to a maximum entropy state (*i.e.*, an equilibrium state) provided

$$\left(\frac{\partial^2 S}{\partial E^2}\right)_V < 0,$$

$$\left(\frac{\partial^2 S}{\partial V^2}\right)_E < 0,$$

with a similar pair of inequalities for system *A*¢. The usual estimate W μ *E^fV^f*, giving $S = kf \ln E + kf \ln V + \ldots$, ensures that the above inequalities are satisfied in conventional macroscopic systems. In the maximum entropy state the systems *A* and *A*¢ have equal temperatures (*i.e.*, they are in thermal equilibrium) and equal pressures (*i.e.*, they are in mechanical equilibrium). The second law of thermodynamics implies that the two interacting systems will evolve towards this state, and will then remain in it indefinitely (if left undisturbed).

RATIO OF THE ENTROPY AND PROBABILITY

The ratio of the probability that all the gas molecules are evenly distributed between the two tanks to the probability that all the molecules, of their own accord and by random motion, would be in tank V_1 is equal to $(V_2/V_1)^N$, where N is the number of molecules.

If V_2/V_1 were equal to 2.0, for example, and N were equal to 10, the probability ratio would be 2 to the tenth power, or 1024. For N = 100, the ratio would be approximately 1.27 times ten to the 30th power. It is clear that the random motion of trillions of gas molecules heavily favours a uniform distribution.

let X_1 = the probability of all the gas molecules, after the valve is opened, remaining in the first tank V_1

let X_2 = the probability of all the gas molecules, after the valve is opened, being uniformly distributed in V_2, the volume of both tanks.

From the probability equation, we have:

$$X_2/X_1 = (V_2/V_1)^N.$$

Taking the natural logarithm of both sides, and then multiplying both sides by R, the gas constant:

$$R \ln(X_2/X_1) = RN \ln(V_2/V_1)$$

$$R/N \ln(X_2/X_1) = R \ln(V_2/V_1)$$

Substituting in equation:

$$\Delta S = R/N \ln(X_2/X_1)$$

Equation represents the fundamental relationship between probability and the second law of thermodynamics. It states that the entropy of a gaseous system increases when its molecular distribution changes from a lower probability to a higher probability (X_2 greater than X_1). Based on the belief that the laws of thermodynamics are universal, this equation has been assumed to apply to all systems, not just gaseous. In other words, any entropy change is proportional to the logarithm of the ratio of probabilities. Therefore, for the general case equation can be written:

$$\Delta S = K \ln(X_2/X_1)$$

Where K is a constant depending on the particular change involved. However, individual values of K, X_1, or X_2 are seldom, if ever, known for non-gaseous systems.

As we have seen before, ΔS can be either positive or negative.

When ΔS is negative equation can be written:

$$-\Delta S = -K \ln(X_2/X_1)$$

$$= K \ln(X_1/X_2)$$

Therefore *a system can go from a more probable state (X_2) to a less probable state (X_1), providing S for the system is negative.* In cases where the system interacts with its surroundings, S can be negative *providing* the over-all entropy of the system and its interacting surroundings is positive; the over-all change can be positive if the entropy increase of the surroundings is numerically greater than the entropy decrease of the system. In the case of the formation of the complex molecules characteristic of living organisms, creationists raise the point that when living things decay after death, the process of decay takes place with an increase in entropy.

They also point out, correctly, that a spontaneous change in a system takes place with a high degree of probability. They fail to realise, however, that probability is relative, and a spontaneous change in a system can be reversed, providing the system interacts with its surroundings in such a manner that the entropy increase in the surroundings is more than enough to reverse the system's original entropy increase. The application of energy can reverse a spontaneous, thermodynamically "irreversible" reaction.

Leaves will spontaneously burn (combine with oxygen) to form water and carbon dioxide. The sun's energy, through the process of photosynthesis, will produce leaves from water vapour and carbon dioxide, and form oxygen. If we unplug a refrigerator, heat will flow to the interior from the surroundings; the entropy increase inside the refrigerator will be greater than the entropy decrease in the surroundings, and the net entropy change is positive.

If we plug it in, this spontaneous "irreversible" change is reversed. Due to the input of electrical energy to the compressor, the heat transferred to the surroundings from the condenser coils is greater than the heat extracted from the refrigerator, and the entropy increase of the surroundings is greater than the entropy decrease of the interior, in spite of the fact that the surroundings are at a higher temperature. Here again, the net entropy change is positive, as would be expected for any spontaneous process.

In a similar manner, the application of electrical energy can reverse the spontaneous reaction of hydrogen and oxygen to form water: when a current is passed through a water solution, hydrogen is liberated at one electrode, oxygen at the other. As can easily be confirmed experimentally, agitating water raises its temperature. When water falls freely from a higher elevation to a lower elevation, its energy is changed from potential to kinetic, and finally to heat as it splashes at the end of its fall.

The second law of thermodynamics states that the water will not spontaneously raise itself to its original elevation using the heat produced on splashing as the sole source of energy. To do so would require a heat engine that would convert all of the heat of splashing to mechanical energy. The efficiency of a heat engine is thermodynamically limited by the Carnot cycle, which limits the efficiency of any heat engine to $\Delta T/T$, where ΔT is the temperature increase due to splashing, and T is the absolute temperature. Since ΔT is only a small fraction of T, there is no device that could be constructed which would allow all of the water to spontaneously jump back to its former elevation.

We can, at least in theory, calculate the entropy increase of the water resulting from its irreversible change in falling. In a manner analogous to that used in the previous example, the entropy increase would be equal to the heat generated by splashing agitation, divided by the absolute temperature. If some of the energy of the falling water is extracted by a water wheel, there will be less heat of splashing and hence less entropy increase. A properly designed turbine could extract most of the water's kinetic energy. This is *not* the same thing as trying to utilize the heat of splashing as an energy source for a heat engine to raise the water. In other words, using the energy before it becomes heat is much more efficient than trying to use it after it becomes heat. If a water wheel is connected by shafts, belts, pulleys, etc. to a pump, the pump can raise water from the downstream side of the water wheel to an elevation even higher than that of the upstream reservoir.

Some of the water would spontaneously raise itself to an elevation even higher than original, but the rest of it would end up below the water wheel on the downstream side. While it is not possible for *all* of the water to raise itself to an elevation higher than its initial elevation, it is possible for *some* of the water to spontaneously raise itself to an elevation higher than initial.

As with any other irreversible change, there will be an increase in overall entropy. This means that the entropy increase of the water going over the wheel is greater than the entropy decrease of water pumped up to the higher elevation. This will be confirmed mathematically in the following paragraphs. will stand for the Greek letter gamma, representing unit weight in pounds per cubic foot. An increase in the value of a parametre will be represented by Δ.

Let:

γ = unit weight of water, lbs./cubic foot

Δh = height of reservoir above downstream side, feet

γh = potential energy of water in reservoir

Δh = additional height above reservoir (to which water is pumped)

$h + \Delta h$ = height to which water is pumped, feet above downstream side

w = work of pumping water to higher elevation

q = heat loss due to pump friction and downstream agitation

f = fraction of water pumped to elevation h + h

From the flow equation, energy in = energy out:

$\gamma h = q + w$

Where:

h = the total energy expended by the water in falling from height h.

q = the energy wasted as heat of splashing

w = work expended in pumping the water to a higher elevation

$q = T\Delta S$

$w = f\gamma(h + \Delta h)$

The total energy available, γh, is divided into pump work, $f\gamma(h + \Delta h)$, and energy lost, $T\Delta S$:

$$\gamma h = T\Delta S + f\gamma(h + \Delta h)$$

Rearranging:

$$T\Delta S = \gamma h - f\gamma(h + \Delta h)$$

When no pump work is done, then:

$$T\Delta S' = \gamma h$$

Combining equations, we get:

$$\Delta S' - \Delta S = f\gamma(h + \Delta h)/T$$

In the case where the water falls freely without turning the water wheel or operating the pump:

let q′= heat of splashing,

$$q' = T\Delta S'$$

where $\Delta S'$ is the entropy increase on free fall,

Equation shows that $\Delta S'$ is larger than ΔS, and that the entropy increase due to pump friction and downstream agitation is "backed up" by the even larger entropy increase that takes place when water falls freely. Equation also shows that the lower the value of ΔS, the more efficient the pump, and the greater the value of f, the fraction of water pumped. Creationists assume that a change characterized by a decrease in entropy can not occur under any circumstances. In fact, spontaneous entropy decreases can, and do, occur all the time, providing sufficient energy is available. The fact that the water wheel and pump are man-built contraptions has no bearing on the case: thermodynamics does not concern itself with the detailed description of a system; it deals only with the relationship between initial and final states of a given system (in this case, the water wheel and pump).

A favourite argument of creationists is that the probability of evolution occurring is about the same as the probability that a tornado blowing through a junkyard could form an airplane. They base this argument on their belief that changes in living things have a very low probability and could not occur without "intelligent design" which overcomes the laws of thermodynamics. This represents a fundamental contradiction in which (they say) evolution is inconsistent with thermodynamics because thermodynamics doesn't permit order to spontaneously arise from disorder, *but* creationism (in the guise of intelligent design) doesn't have to be consistent with the laws of thermodynamics.

A simpler analogy to the airplane/junkyard scenario would be the stacking of three blocks neatly on top of each other. To do this, intelligent design is required, but stacking does not violate the laws of thermodynamics. The same relations hold for this activity as for any other activity involving thermodynamical energy changes. It is true that the blocks will not stack themselves, but as far as *thermodynamics* is concerned, all that is required is the energy to pick them up and place them one on top of the other.

Thermodynamics merely correlates the energy relationships in going from state A to state B. If the energy relationships permit, the change may occur. If they don't permit it, the change can not occur.

A ball will not spontaneously leap up from the floor, but if it is dropped, it will spontaneously bounce up from the floor. Whether the ball is lifted by intelligent design or just happens to fall makes no difference. On the other hand, thermodynamics does not rule out the possibility of intelligent design; it is just simply not a factor with respect to the calculation of thermodynamic probability. Considering the earth as a system, any change that is accompanied by an entropy decrease (and hence going back from higher probability to lower probability) is possible as long as sufficient energy is available. The ultimate source of most of that energy, is of course, the sun.

The numerical calculation of entropy changes accompanying physical and chemical changes are very well understood and are the basis of the mathematical determination of free energy, emf characteristics of voltaic cells, equilibrium constants, refrigeration cycles, steam turbine operating parametres, and a host of other parametres. The creationist position would *necessarily discard the entire mathematical framework* of thermodynamics and would provide no basis for the engineering design of turbines, refrigeration units, industrial pumps, etc. It would do away with the well-developed mathematical relationships of physical chemistry, including the effect of temperature and pressure on equilibrium constants and phase changes.

THE CONSERVATION OF ENERGY

The conservation of energy is the bedrock of our understanding of nature. Conservation of energy, however, only tells us that in any process the total energy remains constant, that is . Energy conservation, however, does not specify what can enter and leave the energy conservation equation, only that the sum of energy entering and leaving a system must always be the same. Furthermore, energy conservation does not tell us what are the various forms of energy, and how energy transforms from one form into another. From our study of waves, light, electric and magnetic fields and so on, we found that energy plays a central role in all these phenomena.

However, only the specific study of the various phenomena could unravel the specific forms that energy takes. The concept of energy also does not tell us what fraction of the total energy is kinetic and what fraction is electrical, chemical and so on. A new concept,in addition to energy, is needed for a more complete understanding of physical processes, in particular those involving the *transformation of energy*.

In all the examples we studied, energy could be completely transformed from one form into another. For example, in our study of magnetic fields, we found that we could fully convert mechanical energy into electrical currents by Faraday's Law of Induction. We may well ask: can all forms of energy be fully transformed from one form to another? The answer is unexpectedly complicated. In some cases the answer is "yes", and in other cases the answer is "no". The concept of entropyaddresses this question regarding the transformation of the forms of energy, and tells us how much energy can be converted from one form into another, and in particular, how much energy is available for doing useful work. Heat had defied all attempts to explain it using the concept of force, and, in general using Newtonian mechanics.

The fact that heat is a form of energy presented a great breakthrough in the understanding of heat, and was fundamental to the emergence of energy as one of the most fundamental ideas in physics. It was, however, soon realised that heat was not an ordinary form of energy such as gravitational or electrical energy.

Rather, heat is a unique form of energy in that only a certain fraction of heat energy can be transformed into other forms of energy. It is for understanding the unique problems presented by heat that new concepts such as entropy, which go beyond energy, were developed. The idea of entropy was introduced into physics in the nineteenth century and is essential for understanding the phenomenon of heat. Entropy is derived from the Greek word meaning "transformation content".

Famous names such as Sadi Carnot, R.J. Clausius, Count Rumford, James Joule, Ludwig Boltzmann and so on contributed greatly to the ideas of heat and entropy. Entropy is a concept which does not answer all the questions left unanswered by energy, but it does address a crucial aspect of energy, namely how much energy can be transformed from one form into another. In particular, if a certain amount of heat energy is given, then entropy tells us how much "useful" work can be extracted from heat energy, in other words, what is the maximum possible efficiency of a heat engine.

In general, since energy is conserved, one would think there is no need to be concerned about recklessly wasting energy. But we know from daily life that "useful" energy, or equivalently, energy "available" for use, is a scarce resource. This intuitive understanding that useful energy is a precious resource is explained by the concept of entropy. Entropy is also related to the concept or order and disorder, and a phase transition from say water to gas is permissible only if entropy increases in such a transformation.

For any system, entropy is a physically measurable quantity. Furthermore, it is an experimentally observed fact that for all processes in nature, the entropy of the total system can never decrease. Note this empirical law concerning entropy is weaker than that of energy, since it does not specify, for any physical process, by how much does entropy increase. The concept of entropy has many other applications and vast ramifications in other disciplines.

Only those chemical processes and reactions for which the total entropy of the system does not decrease are allowed by nature. All living entities must maintain their low entropic state to stay alive.

More abstract applications of entropy occur in the physics of black holes, and it was only in 1997 that the entropy of black holes could be calculated from first principles using results from string theory. The concept of entropy also has a key application in information theory. It was shown by Claude Shannon in 1950 that the information content of a message is determined by a suitable application of the idea of entropy to the discipline of information science. The current understanding of entropy is that it is a large scale (macroscopic) manifestation of the atomic nature of matter. For example, one cubic centimetre of air contains about number of atoms. This is an unimaginably large number of particles. Even if under some circumstances we can view these atoms as classical particles, it is futile to apply Newton's laws to such a large collection of particles.

Instead, the best we can do is to acknowledge our total ignorance and assume that all the atoms are moving randomly, that is, the position and velocity of every atom is a random variable.

Furthermore, for a system of atoms that reaches a stable state, called equilibrium (more on this concept later),every possible velocity and position of the atoms in one cubic centimetre of air is equally likely. \We will show later that the empirical law that entropy never decreases is simply a statement that for a system consisting of a collection of atoms which is not in equilibrium, the collection of atoms will always move towards a state that is the more probable, and will only reach equilibrium on arriving at the most probable state available to the system. It should be noted that the field of thermodynamics developed in the eighteenth and nineteenth centuries without the concept of the atomic composition of matter.

It is remarkable that very general results on how bulk matter behaves were obtained without any idea of the microscopic composition of matter. From the point of view of scientific methodology, the progress of thermodynamics shows that there are concepts describing the large scale properties of matter that have a logic of their own, and which later on are seen to match on smoothly to a deeper and more complete understanding of the same phenomenon.

Temperature and Heat

The thermodynamics, which is the branch of physics dealing with heat, because this will give us a specific physical system in which we can precisely define and discuss entropy and related ideas. In particular, we will study the behaviour of of an ideal gas to illustrate the concept of heat, energy and entropy. Most of us have an intuitive understanding of heat, based on having experienced hot and cold objects. The first thing one must realise is that heat is different from temperature.

CONCEPT OF TEMPERATURE

What do we mean by saying that an object is at temperature ? Temperature is a physical property of all physical systems. From the microscopic point of view, the temperature of a physical object is a measure of the average kinetic energy of the atoms that make up that object. If the atoms, on the average, are moving very fast, they have high kinetic energy and hence the object has a high temperature, and vice versa for slowly moving atoms. We can immediately see that since there is no limit on how much momentum a particle can have, there is consequently no upper limit on how high the temperature can be. The concept of a maximum temperature does, however, appear in Big Bang cosmology, but it is so large as to be practically infinite.

On the other hand, there must be a lower limit to temperature, which is reached when all of the atoms that constitute an object are brought to the state of lowest possible energy, called the ground state. It should be noted that at absolute zero temperature, unlike classical systems is which there is no longer any motion, a quantum system in at absolute zero temperature is in its ground state, and continues to have quantum fluctuations.

ZEROTH LAW OF THERMODYNAMICS

Two objects are in thermal equilibrium if they are at the same temperature. The zeroth law is the basis of the measurement of temperature, and of the existence of thermometres. By bringing an object of known temperature into contact with the object whose temperature is being measured, once thermal equilibrium is reached one can conclude that both the temperatures are equal.

Any physical material whose behaviour under changes of temperature is known can be used to measure temperature. For example, if we know how a metal expands on being heated, or how the resistance of a conductor changes, can be used for measuring temperature. There is an absolute scale of temperature for which the lowest temperature is at a temperature called absolute zero. This scale is called Kelvin and denoted by . Let be the temperature of the object in the Kelvin scale, and T_C be its temperature on the ordinary Celsius scale. We then have the following. $T_C = T - 273.15K$

$$T_C = T - 273.15K$$

Hence, for example, 20 °C = 293.15 K. Temperature is a new dimensional quantity in addition to M,L and T. The units of temperature is K and this is also used to signify its dimension. The fact that it is much more difficult to cool than to heat an object is evidenced by the fact that refrigerators were only made in the twentieth century whereas heat furnaces and engines have been in existence for many centuries. The lowest temperature reached to date was achieved in 1992 and is,

$$T = 2 \times 10^{-9} K$$

For every extra decimal point of cooling closer to absolute zero, new experimental techniques are required and are usually based on qualitatively new physical properties on various phenomena.

HEAT

Every object, at temperature T, has an internal, thermal, energy denoted by E, and which can be thought of as being composed of the sum of the kinetic and potential energy of all the atoms which constitute the object. Heat, denoted by Q, is a form of energy, called thermal energy since it is a form arising from temperature.

Heat always refers to a process in which thermal energy is exchanged between objects at different temperatures. Heat is analogous to work in that both heat and work are defined by a process, be it the process of the flow of heat, or the process of a force acting over some distance. Furthermore, neither heat nor work are intrinsic properties of the system and cannot be stored in the system; what can be stored instead is energy, and which is reflected in the total energy of the system. The dimension of heat is that of energy, and the unit of heat is calorie, defined to be the energy required to heat 1 gm of water by 1 °C. Since J is also a unit of energy, we have the following conversion.

$$1 \text{ Calorie} = 4.186 \text{ J}$$

If two objects having different temperatures are brought into contact such that heat flows between them - called thermal contact - then heat will flow from the object at the higher temperature to that which is at lower temperature, until both the objects reach a common temperature.

The final state is said to be anequilibrium state in which there are no more macroscopic changes taking place. A very important concept in thermodynamics is that of the heat bath, which is taken to be a large reservoir maintained at some constant temperature .

An example of a heat bath is a swimming pool maintained at some constant temperature. When an object is brought into thermal contact with the heat bath, it either gains heat if it is cooler, or loses heat if it is hotter, than ; the important point to note is that in the heat flows, the heat bath is taken to be so big that its temperature does not change, regardless of its exchange of heat with the object in question.

Heat flow is taken to be positive if energy flows into the object and adds to the internal energy E, and is taken to be negative if a part of the internal energy flows out of the system.

Let the system's temperature be . Hence, if $T_S > T_E$, the heat Q that flows is positive, and if $T_S > T_E$, then the heat flow is negative. Figure shows this flow of heat.

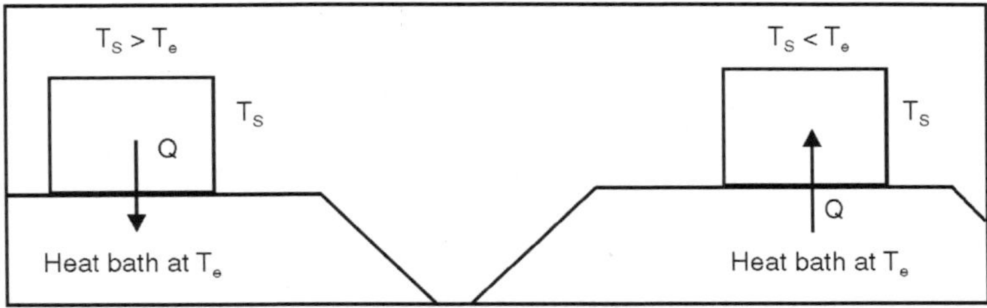

Fig. Heat Flows of Object and Environment

The specific heat of an object, say c, determines how much the temperature of an object, of mass m, changes if it absorbs Q amount of heat. If the initial temperature of an object is T_i and after absorbing Q amount of heat its final temperature is T_f, then the definition of c is given by,

$$T_f - T_i = \frac{Q}{mc}$$

$$\Rightarrow c = \frac{Q}{m(T_F - T_i)}$$

Note heat Q has dimension of energy, that is

$$[Q] = ML^2T^{-2},$$

and hence c has dimension of $[c] = L^2t^{-2}K^{-1}$. Since Q has units of J, the unit c of is $J(Kg)^{-1}K^{-1}$. The specific heat c of an object is a measure of how tightly bound together are its atoms (or molecules). For an object made out of tightly bound constituents, we expect c to be large. A lot of heat has to be supplied to it for increasing its temperature since a larger fraction energy would be stored in the form of potential energy instead of the kinetic energy of moving constituents, and vice versa for materials with loosely bound constituents.

Table. Specific Heat of Some Materials at Room Temperature

Material	$J(Kg)^{-1}K^{-1}$
Lead	128
Silver	236
Copper	386
Aluminum	900
Granite	790
Glass	840
Ice (278K)	2100
Mercury	140
Water	4190
Human Body (average)	3470

So, for example, we can say that metals must have low specific heat since they easily heat up, and this in turn implies that their electrons and atoms are not very tightly bound.

Ideal Gas

Any gas is specified by its temperature T, the volume it occupies V and its pressure P. Pressure is defined to be the force per unit area that the gas exerts on the walls which enclose its volume. T, P, and V are called state variables which specify the state of the gas. Pressure [P] = $ML^{-1}T^{-2}$ and volume [V] = L^3. Note that the state variables are intrinsic properties of the gas, and, unlike heat, do not depend on the history of how the gas arrived at these values. For example, in the phase diagram given above, the state variables say at volume V_0 and temperature T_0 shown above does not depend on the curve which resulted in the gas being at that point of the phase diagram; another process, at a different pressure, leads to the same value of V_0 and T_0, and it does not matter how the gas arrived at that point of phase space.

An ideal gas is a gas that is so dilute that its interactions can be completely neglected, and each atom of the gas can be said to behave as a free particle. A number of gases, including helium, oxygen, nitrogen and so on, do behave as ideal gases at moderate pressures and at moderate temperatures, that is temperatures that are far greater than their condensation temperatures. A derivation is given ofequation of state for the ideal gas based on the kinetic theory. The result is the following. Let N be the number of atoms of a gas in a volume V and the constant be the famous Boltzman's constant K given by,

$$k = 1.38 \times 10^{-23} JK^{-1}$$

We then have the ideal gas law given by,

$$PV = NkT$$

For a gas made out of the collection of molecules (composed of many atoms) the formula above is changed slightly. The state of an ideal gas given by. A phase diagram is one in which any two of its state variables, such as volume and pressure, are plotted as independent variables. A point in the phase diagram represents the state variables of the gas. An implicit assumption in drawing a phase diagram is that thermodynamic variables such as temperature, volume and pressure are continuous variables. This assumption is clearly reasonable, since, in an experiment, we can continuously vary these parametres. The phase diagram for and ideal gas at constant pressure yields a straight line of versus T and is shown in VT phase diagram in Figure . Similarly, the PV phase diagram shows how the pressure P and volume V is related for an ideal gas at different constant temperatures.

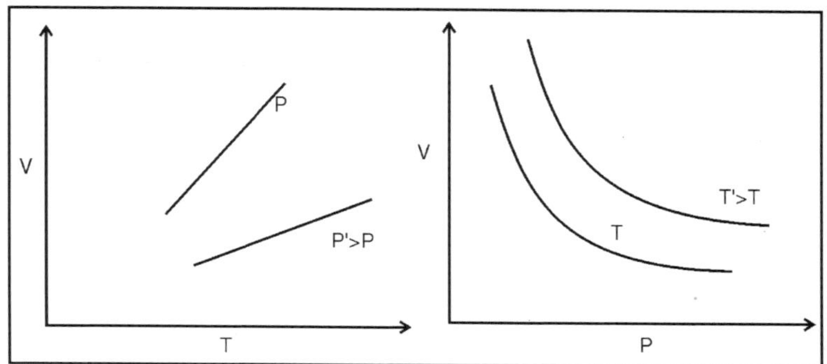

Fig. A VT and a PV Phase Diagram of a Gas

Amazing result. It states that equal volumes of a gas, at a given pressure and temperature, have the same number of atoms, regardless of the kind of atom it is made out of! And this conclusion, as shown, results directly from the application of Newton's Law to a large collection of atoms considered as classical particles. Suppose we hold the temperature of the gas at a constant value, and vary the pressure P. Let the initial values be , and the final values be ,P_f .V_f We then have from that,

$$P_iV_i = P_fV_f\text{: Boyle's Law}$$

Similarly, let us hold the pressure as fixed, and vary the temperature. Let initial values be v_i and T_i, and the final values be V_f and T_f.

Then we have,

$$\frac{V_i}{T_i} = \frac{V_f}{T_f}\text{: Charles's Law}$$

The internal energy U of an ideal gas is given by the following:

$$U = \frac{3}{2}NkT$$

Let us try and understand the physics of the above formulae. For a typical gas, occupying a volume of say 1 litre, the number of particles is close to Avogardo's number, which is given by $N_a = 6.02 \times 10^{23}$. Hence, the number of particles in any volume which we can directly perceive is truly unimaginably large.

The energy per particle is given by,

$$\frac{U}{N} = \frac{3}{2}kT$$

since energy of the particles is purely kinetic, we see that the result is in accordance with our earlier discussion that temperature is a measure of the kinetic energy of an atom in a gas.

The kinetic theory of gases is based on assuming that the position and velocity of the atoms are random. If the random velocity of an atom is v, then, the kinetic energy, , of the gas is simply the average kinetic energy U of the atoms. Denoting the average of value of say v^2 by $< v^2 >$, we have, in three dimensional space.

$$\frac{U}{N} = \frac{1}{2}m < v^2 > = \frac{3}{2}kT$$

COMPLICATIONS WITH FIRST LAW

The main reason that, in the nineteenth century, physics shifted from the idea of force to the idea of energy was due to the need for explaining the phenomenon of heat. If we include heat as a (rather unique) form of energy, the law of energy conservation then goes under the name of the First Law of Thermodyamics, which states the following.

Total Energy of System = Kinetic Energy + Potential Energy + Heat = Constant

Let us examine what form the First Law of Thermodynamics takes for the case of an ideal gas. We would like to apply force on the gas from outside, and determine how much energy and other properties of the gas are changed. We would also like to determine how much work we can extract from a gas. The term work is used for indicating mechanical processes that involve moving around macroscopic bodies, such as pistons and pulleys.

The energy which is expended in doing such work is called mechanical energy, to differentiate it from the energy of the gas, which is called internal energy. Recall work results from a process of a force acting over some distance. The sign convention for work is that work done by the gas is taken to be negative, whereas work done on the gas is taken to be positive.

In other words, a gas loses energy if it does work, whereas it will gain energy if work is done on it. Suppose the gas is contained in a frictionless piston of area , and we hold gas at a fixed temperature by keeping it in contact with a heat bath at temperature T; we do work on the gas by compressing its volume, so that $V_f < V_i$, so that $\Delta V = Vf - Vi < 0$. What is the work that the piston does on the gas? Let the piston exert a force F, and move through a distance Δx; the work done on the gas ΔW is given, as expected, by

$$\Delta W = F\Delta x$$

However, recall that the pressure P exerted by the gas is defined to be force per unit area.

Hence, we have,

$$F = PA$$

and consequently, since the volume is compressed $\Delta V = A\Delta x$, we have,

$$\Delta W = PA\Delta x$$
$$= P\Delta V$$

Suppose that heat of amount ΔQ flows into the gas during compression. We have, from the First Law of Thermody-namics (conservation of energy), the following. Change of Energy of Gas = Work done by gas + Heat flowing into gas,

$$\Delta E = -\Delta W + \Delta Q$$

$$= - P\Delta V + \Delta Q$$

The limit for infinitestimal change is given by,

$$dE = - PdV + dQ$$

Note DE is a function only of the state of the gas, and does not depend on how work was done on, or by, the gas, and how heat flowed in or out of the gas. However, the presence of DQ is process dependent, and we need further analysis to show that E is process independent. Energy conservation does not tell us, for example, how much work can be transformed into heat, or whether all the heat in a body can be made to do "useful" mechanical work. The information about the direction in which energy can flow, and how much energy can be transformed from say heat into work, and vice versa, is given by the concept of entropy.

ENTROPY

A number of concepts before we can precisely define what is the entropy - denoted by - of a physical system. It states that the fundamental result that we will arrive at, namely: For an isolated system, its entropy S can never decrease. The statement above tells us that all physical processes in which heat is transformed into work and visa versa must satisfy the condition that entropy must not decrease.

In other words, if this condition is not satisfied, the process cannot take place. For example a process in which there is a net decrease in entropy, that is, is forbidden as it would imply, that a perpetual machine could be constructed.

To define the concept of entropy quantitatively, we need to first discuss what is meant by a reversible and an irreversible process. Consider a gas inside a frictionless piston in contact with a heat bath. If we pull out the piston very slowly, in a number of very small steps, the gas will expand at constant temperature , and if we then push the piston back, again slowly and in small steps, the gas will compress and return to its former state. This process is reversible as shown in Figure.

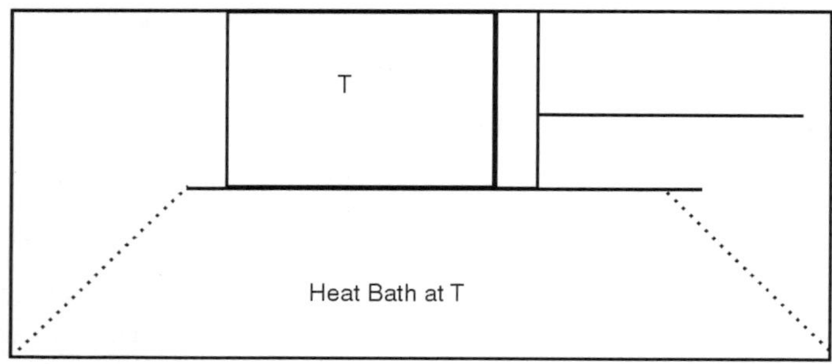

Fig. A Reversible Process

The point to note is that for a reversible process, for each step taken, the system is in equilibrium, and that a small step in one direction, say causing compression, can be exactly reversed by a small step in the opposite direction, causing expansion. If we had suddenly pulled the piston out, the temperature of the gas would have dropped, there would have been turbulent gas motion, heat would have flowed into the gas from the reservoir and so on. The process would then have become an irreversible one. No process in nature is fully reversible, and for a process to be considered as reversible, we need to ignore all effects such as friction, imperfect insulation and so on which would cause irreversiblity. Idealization of a thermodyamic process as being reversible is analogous to Newton's idealization in stating his first law, since complications which come from friction and other irreversible effects are inessential in understanding the underlying principles of thermodyamics.

Consider a system in contact with a heat bath and held at constant temperature T; for a very small change in the independent parametres of the system such as volume, in which ΔQ amount of heat flows into it via a reversible process, the change in entropy of the system is given by,

$$\Delta S = \frac{\Delta Q}{T}$$

Say $[S] = ML^2T^{-2}K^{-1}$,

and the unit of entropy is JK^{-1}. For energy conservation we consequently have,

$$\Delta E = -\Delta W + \Delta Q$$

$$\Rightarrow \Delta E = -P\Delta V + T\Delta S$$

The entropy of an ideal gas is given by,

$$SG = Nk\ \mathrm{In}(VT^{3/2}) + \text{Constant}$$

The entropy per particle (ignoring a constant) is hence given by,

$$S_G / N = k\,\mathrm{In}(VT^{\frac{3}{2}})\cdot$$

We had assumed that for an ideal gas the interactions amongst the particles could be ignored. We see that the formula above confirms this, since the energy and entropy are both proportional to , which comes from a simple sum over the individual gas atoms. Although entropy is an intrinsic property of the system, it is not an intrinsic property of an individual particle, but rather depends, for example, on the total volume occupied by the gas.

To know what a particle is doing, we need to know both its position and velocity. We see that entropy is, in a precise sense, our degree of ignorance about the particle's behaviour. For example, entropy increases as we increase , which reflects the fact that the position of the particle becomes more uncertain, since it can be anywhere in a larger volume .

Similarly, if we increase temperature, entropy goes up since the larger the kinetic energy, the more random the velocity of the particle, and hence a greater degree in our ignorance about the particle.

SIGNIFICANT AND REMARKABLE FEATURE OF ENTROPY

REVERSIBLE PROCESSES

The most significant and remarkable feature of entropy is that, although ΔQ is a quantity which is process dependent, entropy -similar to energy- is anintrinsic property of the system, and unlike heat, entropy can be stored. We hence have S = S(T,V) We examine what it means, for example, for energy and entropy to depend on only T and V.

Consider the VT-phase diagram. That is, consider T and V to be the independent variables, and let pressure P = P(T,V) be fixed by the ideal gas law. For every point on this VT -diagram, both E and S have some definite value. Suppose we take the ideal gas through any process, be it reversible or irreversible, along path 1, as indicated in Figure from values T,V to some new values T′, V′. (Note this path is not a path in physical space, but rather a path in the phase space of T,V.) The system will go through a series of states with varying T,V, each having its own E,S.

The final change in energy and entropy is given by,

$$\Delta E = E(T', V') - E(T, V)$$
$$\Delta S = S(T', V') - S(T, V)$$

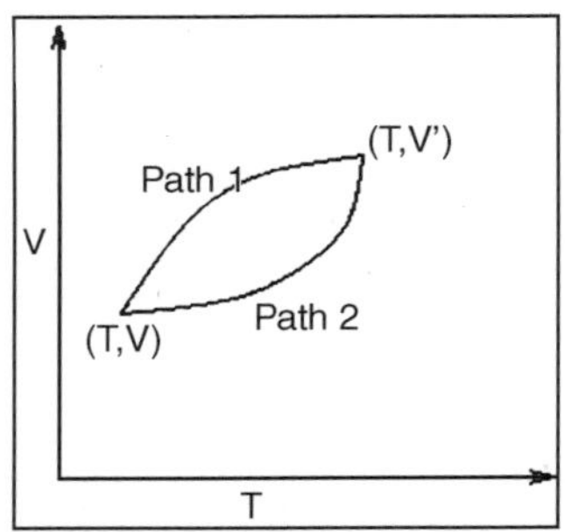

Fig. Phase Diagram of VT with Paths 1 and 2

Now let us take the ideal gas through a different and arbitrary process following say path 2, also given in Figure Then ΔE and ΔS will have the same values as given in! In other words, as mentioned before, the values of E,S depend only on T,V, and not how the system arrived at these values. The change in the energy and entropy for a reversible process does not depend on the path taken; rather it only depends on the initial and final values of entropy. If one goes through a closed path following a reversible process, it follows from above that the total change in the entropy of the system is zero, that is it is restored to its original state. However, note that in this reversible process, a net amount of heat has been expended. In symbols,

$$\Delta_S \text{ Closed Path } = 0 : \text{Reversible}$$
$$\Delta_Q \text{ Closed Path } = >0$$

Recall entropy depends only on the the state of the gas. Hence, if heat ΔQ is supplied to the gas through a reversible process, the total heat expended can be computed from the change in entropy caused by the heat flows. Let the reversible process start with initial state of the gas being Ti, Vi and the final state of the gas being Tf, Vf as shown in figure. The total heat expended in the reversible process is,

$$Q_{Total} = \int_{Reversible} TDS$$

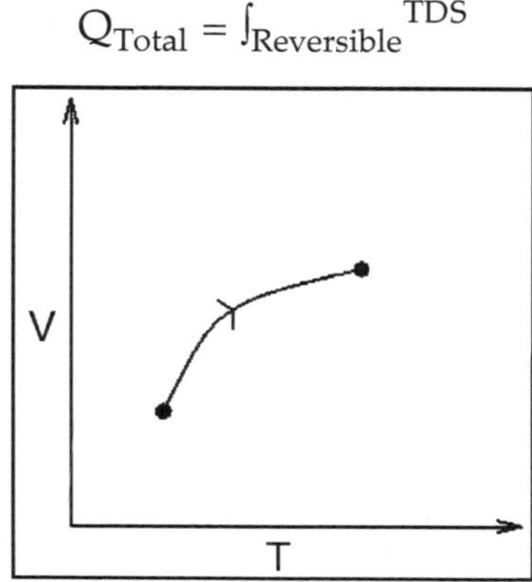

Fig. Reversible Process from Initial State to Final State

IRREVERSIBLE PROCESSES

Consider going around, at room temperature T, in a circle in physical space. Since we always have to overcome friction for movement, certain amount of work W would be necessary to overcome friction. From energy conservation this work would generate heat Q = W and lead to an increase of entropy,

$$\Delta_S = \frac{W}{T} > 0\,.$$

Hence, in going around a closed path, we have not come back to the original state, since we have had heating due to friction, something that we can no longer reverse. In general, if we follow an irreversible process in completing a complete cycle, the entropy always increases.

That is,

$$\Delta_S \text{ Closed Path} > 0 : \text{Irreversible}$$

$$\Delta_Q \text{ Closed Path} > 0$$

How would one determine the change in entropy for an irreversible process? Since the formula we have for entropy given in equation is valid only for reversible processes, we cannot use it. We instead adopt the following approach. The irreversible process starts from some initial state of the system and ends up in a final state. We construct an equivalent reversible process which also goes from the same initial to the same final state. Since entropy does not depend on how its values are attained, we can now compute the change of entropy for the equivalent reversible process, which will then be equal to the change in entropy for the irreversible process in question.

AN IRREVERSIBLE PROCESS: FREE EXPANSION

To illustrate how to calculate the change in entropy in an irreversible process, consider the irreversible free expansion of an ideal gas from an initial volume V_i into a vacuum such that its final volume is V_f. Suppose that the gas is thermally insulated from the environment, and at initial temperature of T_i.

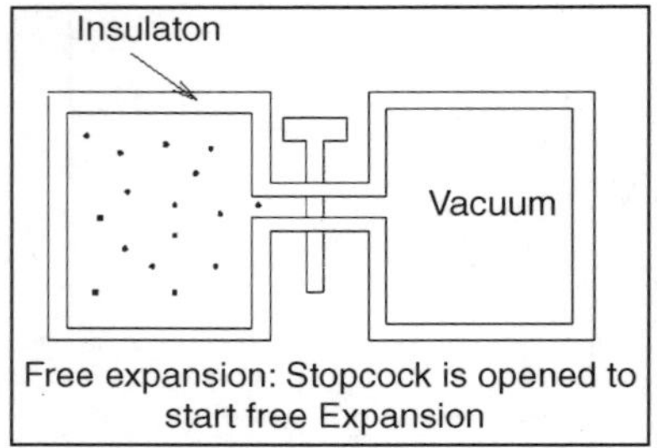

Fig. An Irreversible Process

Since the gas expands into a vacuum it does no work, that is $\Delta W = 0$; furthermore, since it is thermally insulated there is no flow of heat, and hence $\Delta Q = 0$.

The first law implies that,

$$\begin{aligned} \Delta_E &= -\Delta W + \Delta Q \\ &= 0 \end{aligned}$$

Since total energy does not change for an ideal gas, this implies that the initial and final temperature T_f are equal.

Hence,

$$T_i = T_f = T$$

So the initial state of the gas is $T_i = T$, V_i and the final state of the gas is $T_f = T$, V_f. We would like to replace the irreversible free expansion of the ideal gas with an equivalent reversible process. Since the initial and final temperatures are the same, we can replace the free expansion with an (isothermal) expansion of the gas at constant temperature, from volume V_i to V_f. To see that expansion at constant temperature T is a reversible process, consider a gas in a piston and in contact with a heat bath at temperature T.

If one pulls the piston very slowly, heat will flow in to keep the temperature at ; on the other hand, if one slowly compresses the piston, the gas will be compressed, and heat will flow back into the reservoir. So we can see that an isothermal expansion is reversible. All we need to do now is to compute the difference between the entropy of an ideal gas in the initial state T_i, Vi and final state T_f, V_f. We hence have the total change in the entropy as given, from, by,

$$\Delta_S = S_G(T, V_i) - S_G(T, V_f)$$

$$= Nk\ In\left(\frac{V_f}{V_i}\right) > 0 : \text{Free Expansion}$$

CHANGE OF ENTROPY OF GASES

ENTROPY CHANGES IN TERMS OF PROPERTIES OF A PERFECT GAS

From the definition of entropy i.e δQ_{rev} = TδS and from the first law of thermodynamics is dQ = dE + dW which, for a reversible process can be written as δQ_{rev} =δE + PδV... it can be deduced that dE = TdS - PdV for a reversible process and if the substance is a gas of mass m then,

$$dQ = TdS = mC_v dT + PdV \text{...........equation x}$$

For a perfect gas PV = mRT and assuming C_v is constant.

a) Entropy change in terms of V and T

By substituting,

$$P = \frac{RmT}{V}$$

in equation x,

$$\int_1^2 dS = mC_v \int_1^2 \frac{dT}{T} + Rm \int_1^2 \frac{dV}{V} \quad \text{or}$$

$$S_2 - S_1 = mC_v \log_e\left(\frac{T_2}{T_1}\right) + Rm \log_e\left(\frac{V_2}{V_1}\right)$$

b) Entropy change in terms of P and T

By differentiating PV = mRT,

$$PdV = RmdT - VdP = RmdT - \frac{RmT}{P} dP$$

Substituting above in × and integrating to get,

$$\int_1^2 dS = m(C_v + R) \int_1^2 \frac{dT}{T} - Rm \int_1^2 \frac{dP}{P} \quad \text{or}$$

$$S_2 - S_1 = mC_P \log_e\left(\frac{T_2}{T_1}\right) - Rm \log_e\left(\frac{P_2}{P_1}\right)$$

c) Entropy change in terms of P and V,

$$T = \frac{PV}{Rm}$$

and differentiating PV = mRT

Result in,

$$dT = \frac{PdV + VdP}{Rm}$$

Substituting above in × and integrating to get,

$$\int_1^2 dS = m(C_v + R)\int_1^2 \frac{dV}{V} - mC_v \int_1^2 \frac{dP}{P} \quad \text{or}$$

$$S_2 - S_1 = mC_P \log_e\left(\frac{V_2}{V_1}\right) - mC_v \log_e\left(\frac{P_2}{P_1}\right)$$

Example

To determine the entropy change in the irreversible adiabetic expansion of Joules experiment.

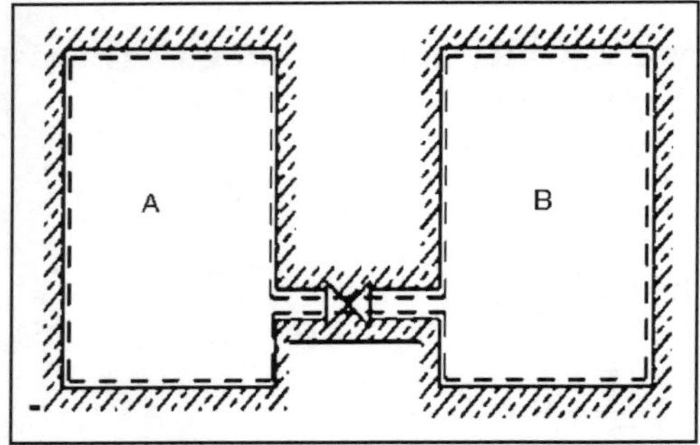

In this experiment a gas at high pressure (properties P_1, V_1,T)in chamber A which is connected to a second evacuated chamber B via a valve. When the valve is opened the gas in chamber A moves by free unresisted expansion into chamber B until the pressures equalise (final properties P_2, V_2,T). The gas is supposed perfect and the temperature does not change (no work has been done). An irreversible expansion has taken place. The equipment is so arranged that there is no heat transfer between the surroundings and the system ($\delta Q = 0$). Although $\delta Q = 0$ the process is irreversible and therefore δQ is not zero. To determine the change in entropy the process is replaced by a reversible process which has the same initial and final conditions *i.e.* an isothermal process.

For an isothermal process the change in entropy,

$$= S_2 - S_1 = mR \log_e \frac{V_2}{V_1}$$

In this case although the process is adiabetic because it is not reversible it is not isentropic. There is a resulting increase in entropy reflecting the various irreversible energy losses in this process.

ENTROPY CHANGE FOR IDEAL FIXED MASS

For an ideal constant volume process... ($dQ = mc_v dT$)

$$dS = \frac{dQ}{T} = \frac{mc_v dT}{T}$$

Integrating between initial and final state,

$$S_2 - S_1 = mc_v \log_e \frac{T_2}{T_1}$$

For an ideal constant pressure process... ($dQ = mc_p dT$),

$$dS = \frac{dQ}{T} = \frac{mc_p dT}{T}$$

Integrating between initial and final state,

$$S_2 - S_1 = mc_v \log_e \frac{T_2}{T_1}$$

For an ideal adiabatic process...

$$dQ = 0$$

$$dS = \frac{dQ}{T} = 0$$

For an ideal isothermal process...

There is no change in temperature and therefore there is no change in internal energy U. From the first law of thermodynamic dQ = dU + dW. If dU = 0 then dQ = dW.

$$dS = \frac{dW}{T} = \frac{pdV}{T}$$

$$\text{as } pV = mRT \text{ or } T = \frac{pV}{mR}$$

$$dS = \frac{pdvmR}{pV} = \frac{mRdV}{V}$$

ADIABETIC,ISENTROPIC, AND ISOTHERMAL PROCESSES

Ideal reversible adiabetic and isothermal processes are straight lines on the TS chart as shown above ac = ideal isothermal process and ab = ideal isentropic process.

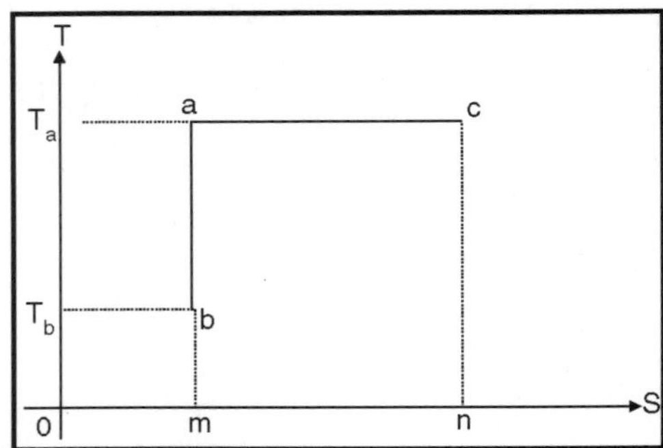

An adiabetic process is one which is insulated against heat transfer and an isentropic process is one with a constant entropy. δQ may be zero during a adiabetic process but if the process is not reversible then δS = δQ/ T does not apply and δS is not zero. If the process is reversible then δQ = 0 and δS = 0 i.e the process is adiabetic and isentropic. A isentropic process is not necessarily reversible for a real expansion or compression-with all the associated eddies and friction associated with real processes may have sufficient heat transfer to maintain the entropy of the relevant substance constant (δS = 0). A real process could be isentropic but would not at the same time be adiabetic because some heat transfer would be necessary. The following relationships are possible

- In a real (irreversible) adiabetic process dQ = 0 but dS > 0
- In a reversible adiabetic process dQ = 0 and dS =0 0
- In a real (irreversible) isentropic process dS = 0 but dQ < 0
- In a reversible isentropic process dS = 0 and dQ = 0 0

Examples of Entropy Changes

Below we have various examples of entropy change during various processes. From them we can draw some general conclusions. First we see in general the entropy of anything increases when it is heated, and the entropy of a gas increases when it expands at constant temperature. A common theme (to be explored in more detail later) is that an increase in entropy is associated with an increase in disorder.

Examples of spontaneous processes are the flow of heat from a hotter to a colder body and the free expansion of a gas. In both of these, the total entropy increases (though that of parts of the system may decrease). This is completely general: spontaneous processes are those which increase entropy. Since heat flow from hotter to colder bodies is irreversible, reversible processes must involve heat flow between bodies of the *same* temperature. It follows that any entropy change of the system must be exactly balanced by that of the heat bath which provided the heat: . Thus the entropy change of the universe during reversible processes is zero. During an adiabatic process no heat flows. Thus from we see that the entropy change of a system during for a reversible, adiabatic process is zero. But note that both qualifiers are needed; the entropy of a non-isolated system can change during a reversible process (and the entropy change of the surroundings will compensate), and an irreversible change to an isolated system will increase the entropy.

Try these problems for yourself before checking the detailed answers!

- *Example.* Two identical blocks of iron, one at 100 °C and the other at 0 °C, are brought into thermal contact. What is the maximum work that can be extracted from the hot block in the absence of other heat sinks?
- *Answer:* The final temperature is 46 °C and 14 per cent of the energy lost by the hot block is available to do work.

Two identical blocks of iron, one at 100 °C and the other at 0 °C, are brought into thermal contact. what is the maximum work that can be extracted from the hot block in the absence of other heat sinks?

Remember, we can't just extract heat from the hot block and turn it into work; the entropy of the block would decrease without any compensating increase elsewhere. We need to add at least enough heat to the cold block so that its entropy increases by as much as that of the hot block decreases. (Work can always be used to do things which don't increase the entropy of the universe, such as lifting a weight.) Once the two blocks are at the same temperature, no further work can be extracted. At that point the entropy change of the two blocks together, from the previous example, is,

$$\Delta S = c \ \text{In}\left(\frac{T_f^2}{T_H T_C}\right) \geq 0$$

The lowest final temperature is that for which , *ie*

$$T_f = \sqrt{T_H T_C} = 319 \text{ K} = 46 \text{ °C}.$$

(Any lower and ΔS would go negative, which isn't allowed.) So although the total heat loss of the hot block is $Q_H = c(T_H - T_f)$, the work extracted is only the difference between this and the heat gained by the cold block,$Q_C = c(T_f - T_C)$, namely $W = c(T_H + T_C - 2T_f)$.

In this case, the efficiency is,

$$\eta = \frac{W}{Q_H} = \frac{(T_H + T_C - 2T_f)}{(T_H - T_f)} = 0.144$$

Example. Heat engines revisited From the law of non-decrease of entropy, show that the maximum efficiency of a heat engine operating between two reservoirs at and occurs when the engine is reversible.

Answer: Heat engines revisited: From the law of non-decrease of entropy, show that the maximum efficiency of a heat engine operating between two reservoirs at T_H and T_C occurs when the engine is reversible. This is a bit circular, as we used the properties of Carnot engines to derive this form of the second law! However we will see later in the course that it can be independently derived using statistical methods. The change in entropy of the two reservoirs, which must be non-negative, is

$$\Delta S = \Delta S_C + \Delta S_H = \frac{Q_C}{T_C} - \frac{Q_H}{T_H} \geq 0 \Rightarrow \frac{Q_C}{Q_H} \geq \frac{T_C}{T_H}$$

with the equality being for $\Delta S = 0$, *ie* for a reversible process. So the efficiency is,

$$\eta = \frac{W}{Q_H} = 1 - \frac{Q_C}{Q_H} \leq 1 - \frac{T_C}{T_H}$$

This is maximum when the equality is satisfied, *i.e.*, for a reversible engine.

Example. n moles of an ideal gas at temperature T_0 are originally confined to half of an insulated container by a partition. The partition is removed without doing any work. What is the change in entropy?

Answer: ΔS = nRIn2. More details n moles of an ideal gas at temperature T_0 are originally confined to half of an insulated container by a partition. The partition is removed without doing any work. What is the final change in entropy?

First, we must resist the temptation to say $đQ = 0$ (because process is adiabatic) hence dS = 0. In fact $dS = đQ^{rev}/T$ and this is not a reversible process. To solve this problem, we need to find a reversible process linking the same two endpoints. We've already seen this; the process is a reversible isothermal expansion

For such an expansion,

$$Q = -W = nRT_0 \text{ In}(V_2/V_1) \text{ so } \Delta S = Q/T_0 = nR \text{ In}2.$$

This is rather subtle; make sure you understand the difference between the actual process (with Q = 0 and W = 0) and the reversible process for which we could calculate the entropy change (with $Q = -W \neq 0$). Remember, Q and W are not functions of state - only the sum is.

Example:.An insulated container is originally divided in half by a partition, and each half is originally occupied by n moles of an different ideal gas at temperature T_0. The partition is removed without doing any work. What is the change in entropy?

Answer: ΔS = 2nR In2. An insulated container is originally divided in half by a partition, and each half is originally occupied by moles of an different ideal gas at temperature . The partition is removed without doing any work. What is the change in entropy?

Since these are ideal gases, they do not interact. Each species is oblivious of the existence of the other and the entropy change for each is just the same as in the free expansion of . Thus the total entropy change is given by . Note that the total mixing - which we know will happen eventually - is exactly the change which maximises the entropy.

Example. Two identical blocks of iron, one at 100 °C and the other at 0 °C, are brought into thermal contact. What happens? What is the total entropy change? (Assume the heat capacity of each block, C, is constant over this temperature range, and neglect volume changes)

Answer: Both blocks end up at 50 °C and the entropy change is 0.024 °C.

Two identical blocks of iron, one at 100 °C and the other at 0 °C, are brought into thermal contact. What happens? What is the total entropy change? (Assume the heat capacity of each block, , is constant over this temperature range, and neglect volume changes)

We know what happens: Heat flows from the hot to the cold body till they reach the same temperature; conservation of energy requires that this will be at 50 °C. Why does heat transfer occur? If heat $đQ$ is transferred from a hot body at T_H to a cold one at T_C, the entropy decrease of the hot body is $dS_H > -đQ / T_H$, and the entropy increase of the cold body is $dS_C > đQ / T_C$. So the total entropy change is,

$$dS = dS_H + dS_C > \left(\frac{1}{T_C} - \frac{1}{T_H} \right) đQ > 0$$

So the decrease in entropy of the hot block is more than compensated by the increase in entropy of the cold block. The spontaneous flow of heat is associated with an overall entropy increase, and the two blocks exchange heat till their combined entropy is maximised.

What is the overall entropy change for the total process? Here we have to use a trick. We can't calculate entropy changes for irreversible processes directly; we need to imagine a reversible process between the same endpoints. For the heating or cooling of a block, this would involve bringing it in contact with a series of heat baths at infinitesimally increasing or decreasing temperatures, so that the temperature difference between the heat bath and the block is always negligible and the entropy change is zero.

During this process, the heat transfer and the infinitesimal temperature change are related by,

$$đQ^{rev} = CdT$$

$$\Delta S = \int_1^2 \frac{đQ^{rev}}{T} = C \int_{T_1}^{T_2} \frac{dT}{T} = C \operatorname{In} \left(\frac{T_2}{T_1} \right)$$

Thus the total entropy change is,

$$\Delta S = \Delta S_C + \Delta S_H = C \text{ In}\left(\frac{T_f}{T_C}\right) + C \text{ In}\left(\frac{T_f}{T_H}\right) = C \text{ In}\left(\frac{T_f^2}{T_H T_C}\right)$$

$$= C \text{ In}\left(\frac{323^2}{273 \times 373}\right) = 0.024C$$

ENTROPY IN QUANTUM MECHANICS

In quantum statistical mechanics, the concept of entropy was developed by John von Neumann and is generally referred to as "von Neumann entropy". Von Neumann established a rigorous mathematical framework for quantum mechanics with his work Mathematische Grundlagen der Quantenmechanik.

He provided in this work a theory of measurement, where the usual notion of wave collapse is described as an irreversible process (the so called von Neumann or projective measurement).

Using this concept, in conjunction with the density matrix he extended the classical concept of entropy into the quantum domain.

It is well known that a Shannon based definition of information entropy leads in the classical case to the Boltzmann entropy.

It is tempting to regard the Von Neumann entropy as the corresponding quantum mechanical definition. But the latter is problematic from quantum information point of view. Consequently Stotland, Pomeransky, Bachmat and Cohen have introduced a new definition of entropy that reflects the inherent uncertainty of quantum mechanical states.

This definition allows to distinguish between the minimum uncertainty entropy of pure states, and the excess statistical entropy of mixtures.

Entropy in Astrophysics

In astrophysics, what is referred to as "entropy" is actually the adiabatic constant derived as follows. Using the first law of thermodynamics for a quasi-static, infinitesimal process for a hydrostatic system,

$$dQ = dU - dW.$$

For an ideal gas in this special case, the internal energy, U, is only a function of T; therefore the partial derivative of heat capacity with respect to T is identically the same as the full derivative, yielding through some manipulation,

$$dQ = C_V dT + PdV.$$

Further manipulation using the differential version of the ideal gas law, the previous equation, and assuming constant pressure, one finds,

$$dQ = C_P dT - VdP.$$

For an adiabatic process dQ = 0 and recalling g $= \frac{C_P}{C_V}$, one finds,

$$VdP = C_P dT$$

$$PdV = - C_V dT$$

$$\frac{dP}{P} = -\frac{dV}{P}\gamma.$$

One can solve this simple differential equation to find,

$$PV^{\gamma} = \text{constant} = K$$

This equation is known as an expression for the adiabatic constant, K, also called the adiabat. From the ideal gas equation one also knows

$$P = \frac{\rho k_B T}{\mu m_H},$$

where k_B is Boltzmann's constant. Substituting this into the above equation along with V = [*grams*]/r and g = 5/3 for an ideal monoatomic gas one finds,

$$K = \frac{k_B T}{\mu m_H r^{2/3}},$$

where m is the mean molecular weight of the gas or plasma; and m_H is the mass of the Hydrogen atom, which is extremely close to the mass of the proton, m_p, the quantity more often used in astrophysical theory of galaxy clusters. This is what astrophysicists refer to as "entropy" and has units of [keV cm^2]. This quantity relates to the thermodynamic entropy as

$$S = k_B \ln \Omega + S_0$$

where W, the density of states in statistical theory, takes on the value of K as defined above.

Energy Dispersal

The concept of entropy can be described qualitatively as a measure of energy dispersal at a specific temperature. Similar terms have been in use from early in the history of classical thermodynamics, and with the development of statistical thermodynamics and quantum theory, entropy changes have been described in terms of the mixing or "spreading" of the total energy of each constituent of a system over its particular quantized energy levels.

Ambiguities in the terms disorder and chaos, which usually have meanings directly opposed to equilibrium, contribute to widespread confusion and hamper comprehension of entropy for most students. As the second law of thermodynamics shows, in an isolated system internal portions at different temperatures will tend to adjust to a single uniform temperature and thus produce equilibrium.

A recently developed educational approach avoids ambiguous terms and describes such spreading out of energy as dispersal, which leads to loss of the differentials required for work even though the total energy remains constant in accordance with the first law of thermodynamics. Physical chemist Peter Atkins, for example, who previously wrote of dispersal leading to a disordered state, now writes that "spontaneous changes are always accompanied by a dispersal of energy", and has discarded 'disorder' as a description.

Entropy and Information Theory

In information theory, entropy is the measure of the amount of information that is missing before reception and is sometimes referred to as Shannon entropy. Shannon entropy is a broad and general concept which finds applications in information theory as well as thermodynamics.

It was originally devised by Claude Shannon in 1948 to study the amount of information in a transmitted message. The definition of the information entropy is, however, quite general, and is expressed in terms of a discrete set of probabilities pi. In the case of transmitted messages, these probabilities were the probabilities that a particular message was actually transmitted, and the entropy of the message system was a measure of how much information was in the message.

For the case of equal probabilities (*i.e.* each message is equally probable), the Shannon entropy (in bits) is just the number of yes/no questions needed to determine the content of the message. The question of the link between information entropy and thermodynamic entropy is a hotly debated topic. Some authors argue that there is a link between the two, while others will argue that they have absolutely nothing to do with each other.

The expressions for the two entropies are very similar. The information entropy H for equal probabilities pi = p is:

$$H = K \ln (1/p)$$

where K is a constant which determines the units of entropy. For example, if the units are bits, then K = 1/ln. The thermodynamic entropy S , from a statistical mechanical point of view was first expressed by Boltzmann:

$$S = 1 \ln (1/p)$$

where p is the probability of a system being in a particular microstate, given that it is in a particular macrostate, and k is Boltzmann's constant. It can be seen that one may think of the thermodynamic entropy as Boltzmann's constant, divided by ln, times the number of yes/no questions that must be asked in order to determine the microstate of the system, given that we know the macrostate.

The link between thermodynamic and information entropy was developed in a series of papers by Edwin Jaynes beginning in 1957. The problem with linking thermodynamic entropy to information entropy is that in information entropy the entire body of thermodynamics which deals with the physical nature of entropy is missing.

The second law of thermodynamics which governs the behaviour of thermodynamic systems in equilibrium, and the first law which expresses heat energy as the product of temperature and entropy are physical concepts rather than informational concepts.

If thermodynamic entropy is seen as including all of the physical dynamics of entropy as well as the equilibrium statistical aspects, then information entropy gives only part of the description of thermodynamic entropy.

Some authors, like Tom Schneider, argue for dropping the word entropy for the H function of information theory and using Shannon's other term "uncertainty" instead.

Ice Melting Example

The illustration for this article is a classic example in which entropy increases in a small 'universe', a thermodynamic system consisting of the 'surroundings' (the warm room) and 'system' (glass, ice, cold water). In this universe, some heat energy dQ from the warmer room surroundings (at 298 K or 25 °C) will spread out to the cooler system of ice and water at its constant temperature T of 273 K (0 °C), the melting temperature of ice.

The entropy of the system will change by the amount $dS = dQ/T$, in this example dQ/273 K. (The heat dQ for this process is the energy required to change water from the solid state to the liquid state, and is called the enthalpy of fusion, *i.e.* the DH for ice fusion.) The entropy of the surroundings will change by an amount $dS = -dQ/298$ K. So in this example, the entropy of the system increases, whereas the entropy of the surroundings decreases.

It is important to realise that the decrease in the entropy of the surrounding room is less than the increase in the entropy of the ice and water: the room temperature of 298 K is larger than 273 K and therefore the ratio, (entropy change), of dQ/298 K for the surroundings is smaller than the ratio (entropy change), of dQ/273 K for the ice+water system.

To find the entropy change of our "universe", we add up the entropy changes for its constituents: the surrounding room and the ice +water. The total entropy change is positive; this is always true in spontaneous events in a thermodynamic system and it shows the predictive importance of entropy: the final net entropy after such an event is always greater than was the initial entropy.

As the temperature of the cool water rises to that of the room and the room further cools imperceptibly, the sum of the dQ/T over the continuous range, at many increments, in the initially cool to finally warm water can be found by calculus. The entire miniature "universe", *i.e.* this thermodynamic system, has increased in entropy. Energy has spontaneously become more dispersed and spread out in that "universe" than when the glass of ice water was introduced and became a "system" within it.

Entropy and Cosmology

As a finite universe may be considered an isolated system, it may be subject to the Second Law of Thermodynamics, so that its total entropy is constantly increasing.

It has been speculated that the universe is fated to a heat death in which all the energy ends up as a homogeneous distribution of thermal energy, so that no more work can be extracted from any source. If the universe can be considered to have generally increasing entropy, then – as Roger Penrose has pointed out – gravity plays an important role in the increase because gravity causes dispersed matter to accumulate into stars, which collapse eventually into black holes.

Jacob Bekenstein and Stephen Hawking have shown that black holes have the maximum possible entropy of any object of equal size. This makes them likely end points of all entropy-increasing processes, if they are totally effective matter and energy traps. Hawking has, however, recently changed his stance on this aspect.

The role of entropy in cosmology remains a controversial subject. Recent work has cast extensive doubt on the heat death hypothesis and the applicability of any simple thermodynamic model to the universe in general.

Although entropy does increase in the model of an expanding universe, the maximum possible entropy rises much more rapidly – thus entropy density is decreasing with time. This results in an "entropy gap" pushing the system further away from equilibrium. Other complicating factors, such as the energy density of the vacuum and macroscopic quantum effects, are difficult to reconcile with thermo-dynamical models, making any predictions of large-scale thermodynamics extremely difficult.

MEASUREMENT OF ENTROPY

In real experiments, it is quite difficult to measure the entropy of a system. The techniques for doing so are based on the thermodynamic definition of the entropy, and require extremely careful calorimetry.

For simplicity, we will examine a mechanical system, whose thermodynamic state may be specified by its volume V and pressure P. In order to measure the entropy of a specific state, we must first measure the heat capacity at constant volume and at constant pressure (denoted C_V and C_P respectively), for a successive set of states intermediate between a reference state and the desired state. The heat capacities are related to the entropy S and the temperature T by

$$C_X = T\left(\frac{\partial S}{\partial T}\right)_X$$

where the X subscript refers to either constant volume or constant pressure. This may be integrated numerically to obtain a change in entropy:

$$\Delta S = \int \frac{C_X}{T} dT$$

We can thus obtain the entropy of any state (P,V) with respect to a reference state (P_0, V_0). The exact formula depends on our choice of intermediate states. For example, if the reference state has the same pressure as the final state,

$$S(P,V) = S(P,V_0) + \int_{T(P,V_0)}^{T(P,V)} \frac{C_P(P,V(T,P))}{T} dT$$

In addition, if the path between the reference and final states lies across any first order phase transition, the latent heat associated with the transition must be taken into account.

The entropy of the reference state must be determined independently. Ideally, one chooses a reference state at an extremely high temperature, at which the system exists as a gas. The entropy in such a state would be that of a classical ideal gas plus contributions from molecular rotations and vibrations, which may be determined spectroscopically. Choosing a *low* temperature reference state is sometimes problematic since the entropy at low temperatures may behave in unexpected ways. For instance, a calculation of the entropy of ice by the latter method, assuming no entropy at zero temperature, falls short of the value obtained with a high-temperature reference state by 3.41 J/(mol·K). This is due to the "zero-point" entropy of ice mentioned earlier.

AVAILABLE USEFUL WORK

An important and revealing idealised special case is to consider applying the Second Law to the scenario of an isolated system (called the total system or universe), made up of two parts: a sub-system of interest, and the sub-system's surroundings. These surroundings are imagined to be so large that they can be considered as an *unlimited* heat reservoir at temperature T_R and pressure P_R — so that no matter how much heat is transferred to (or from) the sub-system, the temperature of the surroundings will remain T_R; and no matter how much the volume of the sub-system expands (or contracts), the pressure of the surroundings will remain P_R.

Whatever changes dS and dS_R occur in the entropies of the sub-system and the surroundings individually, according to the Second Law the entropy S_{tot} of the isolated total system must increase:

$$dS_{\text{tot}} = dS + dS_R \geq 0$$

According to the First Law of Thermodynamics, the change dU in the internal energy of the sub-system is the sum of the heat dq added to the sub-system, *less* any work dw done *by* the sub-system, *plus* any net chemical energy entering the sub-system $\Sigma\mu_{iR}N_i$, so that:

$$dU = dq - dw + d\,(Sm_{iR}N_i)$$

where m_{iR} are the chemical potentials of chemical species in the external surroundings.

Now the heat leaving the reservoir and entering the sub-system is

$$dq = T_R(-dS_R) \leq T_R dS$$

where we have first used the definition of entropy in classical thermodynamics (alternatively, the definition of temperature in statistical thermodynamics); and then the Second Law inequality from above.

It therefore follows that any net work dw done by the sub-system must obey

$$dw \leq -dU + T_R dS + Sm_{iR}dN_i$$

It is useful to separate the work done dw done by the subsystem into the *useful* work dw_u that can be done *by* the sub-system, over and beyond the work $P_R\ dV$ done merely by the sub-system expanding against the surrounding external pressure, giving the following relation for the useful work that can be done:

$$dw_u \leq -d(U - T_R S + P_R V - Sm_{iR}N_i)$$

It is convenient to define the right-hand-side as the exact derivative of a thermodynamic potential, called the *availability* or exergy X of the subsystem,

$$X = U - T_R S + P_R V - Sm_{iR}N_i$$

The Second Law therefore implies that for any process which can be considered as divided simply into a subsystem, and an unlimited temperature and pressure reservoir with which it is in contact,

$$dX + dw_u \leq 0$$

i.e. the change in the subsystem's exergy plus the useful work done *by* the subsystem (or, the change in the subsystem's exergy less any work, additional to that done by the pressure reservoir, done *on* the system) must be less than or equal to zero.

VARIOUS PROCESSE OF ENTROPY CHANGE

CYCLIC PROCESS

When a system in a given initial state goes through a number of different changes in state (going through various processes) and finally returns to its initial values, the system has undergone a cyclic process or cycle. Therefore, at the conclusion of a cycle, all the properties have the same value they had at the beginning. Steam (water) that circulates through a closed cooling loop undergoes a cycle.

REVERSIBLE PROCESS

A reversible process for a system is defined as a process that, once having taken place, can be reversed, and in so doing leaves no change in either the system or surroundings. In other words the system and surroundings are returned to their original condition before the process took place. In reality, there are no truly reversible processes; however, for analysis purposes, one uses reversible to make the analysis simpler, and to determine maximum theoretical efficiencies.

Therefore, the reversible process is an appropriate starting point on which to base engineering study and calculation. Although the reversible process can be approximated, it can never be matched by real processes. One way to make real processes approximate reversible process is to carry out the process in a series of small or infinitesimal steps. For example, heat transfer may be considered reversible if it occurs due to a small temperature difference between the system and its surroundings. For example, transferring heat across a temperature difference of 0.00001 °F "appears" to be more reversible than for transferring heat across a temperature difference of 100 °F.

Therefore, by cooling or heating the system in a number of infinitesamally small steps, we can approximate a reversible process. Although not practical for real processes, this method is beneficial for thermodynamic studies since the rate at which processes occur is not important.

IRREVERSIBLE PROCESS

An irreversible process is a process that cannot return both the system and the surroundings to their original conditions. That is, the system and the surroundings would not return to their original conditions if the process was reversed. For example, an automobile engine does not give back the fuel it took to drive up a hill as it coasts back down the hill. There are many factors that make a process irreversible.,Four of the most common causes of irreversibility are friction, unrestrained expansion of a fluid, heat transfer through a finite temperature difference, and mixing of two different substances. These factors are present in real, irreversible processes and prevent these processes from being reversible.

ADIABATIC PROCESS

An adiabatic process is one in which there is no heat transfer into or out of the system. The system can be considered to be perfectly insulated.

Example:

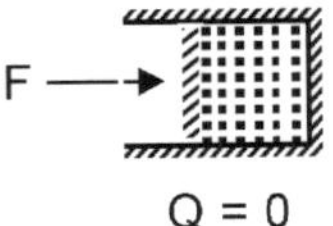

Compression of a Gas in an Insulated Cylinder. In this case any change in the internal energy of the gas is due to work done on it or by it, $\Delta U = W$. Normally if ΔU changes the temperature of a system will change. Any temperature rise or fall is due to the work done or by the gas alone and not due to heat flowing into or out of the system since $Q = 0$. If a process is carried out fast enough the heat flow will be small and the process can be approximate as being adiabatic. This happen because heat flow is in general a slow process. Observe that we did not say that Q is constant because it not a state variable. Q represent an energy transfer not the heat energy of the system. In addition to the ideal gas law $PV = NkT$, the quantity PV^g is constant for an ideal gas where $g = c_P/c_V$, the ratio of molar specific heats.

For an ideal gas the work,

$$W = (P_1V_1 - P_2V_2)/(\gamma - 1)$$

ISOENTROPIC PROCESS

An isoentropic process is one in which the entropy of the fluid remains constant. This will be true if the process the system goes through is reversible and adiabatic. An isoentropic process can also be called a constant entropy process.

Example: A Heat Engine in which the working fluid undergoes an Adiabatic Reversible cyclic process.

T_{high}

Q_2

$\Delta S = 0$ — W

Q_1

T_{low}

Any isoentropic process is also adiabatic since DU = DdQ/T and Q = 0. However, not all adiabatic process are isoentropic. Adiabatic free expansion is not isoentropic. For a heat engine to be reversible, not only must the change in entropy of the working fluid be zero but also the net change of the entropy of the environment (heat reservoirs) must also be zero.

POLYTROPIC PROCESS

When a gas undergoes a reversible process in which there is heat transfer, the process frequently takes place in such a manner that a plot of the Log P (pressure) vs. Log V (volume) is a straight line. Or stated in equation form PV^n = a constant. This type of process is called a polytropic process. An example of a polytropic process is the expansion of the combustion gasses in the cylinder of a water-cooled reciprocating engine.

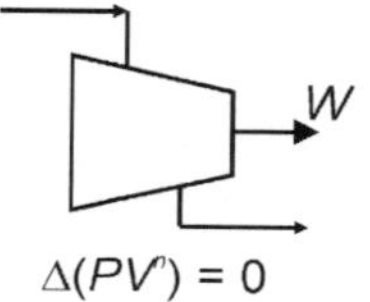

Here n can be any real number. All the process above can thought of as a special case of a polytropic process for an ideal gas.

n = 0 for Isobaric process since $PV^0 = P$ = constant.

n = 1 for Isothermal process since $PV^1 = PV = NKT$ = constant.

n = ∞ for Isovolumetric process

n = **g** for Adiabatic process

THROTTLING PROCESS

A throttling process is defined as a process in which there is no change in enthalpy from state one to state two, $h_1 = h_2$; no work is done, W = 0; and the process is adiabatic, Q = 0. To better understand the theory of the ideal throttling process let's compare what we can observe with the above theoretical assumptions. An example of a throttling process is an ideal gas flowing through a valve in midposition. From experience we can observe that:

$$P_{in} > P_{out}, \; vel_{in} < vel_{out}$$

where,

P = pressure

vel = velocity

These observations confirm the theory that,

$$h_{in} = h_{out}$$

Remember h = u + Pv (v = specific volume), so if pressure decreases then specific volume must increase if enthalpy is to remain constant (assuming u is constant). Because mass flow is constant, the change in specific volume is observed as an increase in gas velocity, and this is verified by our observations. The theory also states W = 0. Our observations again confirm this to be true as clearly no "work" has been done by the throttling process. Finally, the theory states that an ideal throttling process is adiabatic. This cannot clearly be proven by observation since a "real" throttling process is not ideal and will have some heat transfer.

ENTROPY CHANGE FOR IDEALLY CLOSED SYSTEM GAS

For an ideal constant volume process... ($dQ = mc_v dT$)

$$dS = \frac{dQ}{T} = \frac{mc_v dT}{T}$$

Integrating between initial and final state

$$S_2 - S_1 = mc_v \log_\theta \frac{T}{T_1}$$

For an ideal constant pressure process... ($dQ = mc_p dT$)

$$dS = \frac{dQ}{T} = \frac{mc_p dT}{T}$$

Integrating between initial and final state,

$$S_2 - S_1 = mc_p \log_e \frac{T_2}{T_1}$$

For an ideal adiabetic process...

$$dQ = 0$$

$$dS = \frac{dQ}{T} = 0$$

For an ideal isothermal process...

There is no change in temperature and therefore there is no change in internal energy U.

From the first law of thermodynamic dQ = dU + dW. If dU = 0 then dQ = dW.

$$dS = \frac{dW}{T} = \frac{pdV}{T}$$

$$\text{as } pV = mRT \text{ or } T = \frac{pV}{mR}$$

$$dS = \frac{pdVmR}{pV} = \frac{mRdV}{V}$$

Integrating between initial and final state,

$$S_2 - S_1 = mR \log_e \frac{V_2}{V_1}$$

Entropy changes in terms of properties of a perfect gas.

From the definition of entropy i.e dQ_{rev} = TdS and from the first law of thermodynamics is dQ = dE + dW which, for a reversible process can be written as dQ_{rev} =dE + PdV... it can be deduced that dE = TdS - PdV for a reversible process and if the substance is a gas of mass m then

$$dQ = TdS = mc_v dT + PdV \text{........equation X}$$

For a perfect gas PV = mRT and assuming C_v is constant.

- Entropy change in terms of V and T

$$\text{by substituting } P = \frac{RmT}{V} \text{ in equation x}$$

$$\int_1^2 dS = mC_v \int_1^2 \frac{dT}{T} + Rm \int_1^2 \frac{dV}{V}$$

$$S_2 - S_1 = mC_v \log_e \left(\frac{T_2}{T_1}\right) + Rm \log_e \left(\frac{V_2}{V_1}\right)$$

- Entropy change in terms of P and T By differentiating PV = mRT

$$PdV = RmdT - VdP = RmdT - \frac{RmT}{P}dP$$

Substituting above in x and integrating to get

$$\int_1^2 dS = m(C_v + R)\int_1^2 \frac{dT}{T} - Rm\int_1^2 \frac{dP}{P} \text{ or}$$

$$S_2 - S_1 = mC_P \log_e\left(\frac{T_2}{T_1}\right) - Rm\log_e\left(\frac{P_2}{P_1}\right)$$

- Entropy change in terms of P and V

$T = \frac{PV}{Rm}$ and differentiating $PV = mRT$

Result in $dT = \frac{PdV + VdP}{Rm}$

Substituting above in x and integrating to get

$$\int_1^2 dS = m(C_v + R)\int_1^2 \frac{dV}{V} - mC_v\int_1^2 \frac{dP}{P} \text{ or}$$

$$S_2 - S_1 = mC_P \log_e\left(\frac{V_2}{V_1}\right) - mC_v\log_e\left(\frac{P_2}{P_1}\right)$$

5

Analogy between Fluid Friction and Heat Transfer

INTRODUCTION

We describe the physical mechanism for the heat transfer coefficient in a turbulent boundary layer because most aerospace vehicle applications have turbulent boundary layers. The treatment closely follows that in Eckert and Drake. Very near the wall, the fluid motion is smooth and laminar, and molecular conduction and shear are important. The shear stress, τ, at a plane is given by $\mu \frac{dc}{dy} = \tau$ (where μ is the dynamic viscosity), and the heat flux by

$$\dot{q} = -k \frac{dT}{dy}.$$

The latter is the same expression that was used for a solid. The boundary layer is a region in which the velocity is lower than the free stream as shown in Figures. In a turbulent boundary layer, the dominant mechanisms of shear stress and heat transfer change in nature as one moves away from the wall.

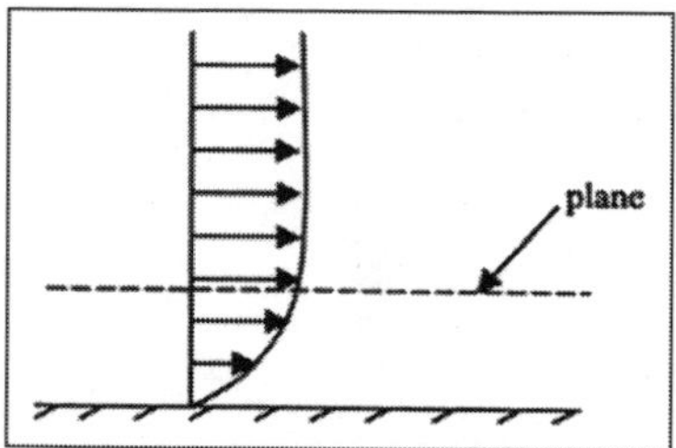

Fig. Velocity Profile Near a Surface

As one moves away from the wall (but still in the boundary layer), the flow is turbulent. The fluid particles move in random directions and the transfer of momentum and energy is mainly through interchange of fluid particles, shown schematically in Figure:

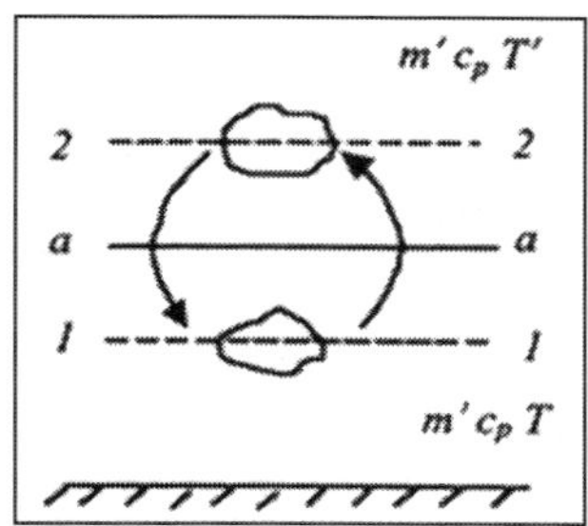

Fig. Momentum and Energy Exchanges in Turbulent Flow

With reference to Figure, because of the turbulent velocity field, a fluid mass m' penetrates the plane *aa* per unit time and unit area. In steady flow, the same amount crosses *aa* from the other side. Fluid moving up transports heat $m'c_pT$. Fluid moving down transports $m'c_pT'$ downwards. If $T' > T$, there is a turbulent downwards heat flow $\dot{q}_{tubulent}$, given by $\dot{q}_{tubulent} = m'c_p(T' - T)$, that results. Fluid moving up also has momentum $m'c$ and fluid moving down has momentum $m'c'$. The net flux of momentum down per unit area and time is therefore $m'(c' - c)$. This net flux of momentum per unit area and time is a force per unit area or stress, given by

$$t_{turbulent} = m'(c' - c)$$

Based on these considerations, the relation between heat flux and shear stress at plane *aa* is

$$\dot{q}_{turbulent} = \tau_{turbulent} c_p \left(\frac{T' - T}{c' - c} \right),$$

or (again approximately),

$$\dot{q}_{turbulent} = \tau_{turbulent} c_p \left(\frac{dT}{dc} \right),$$

since the locations of planes 1–1 and 2–2 are arbitrary.

For the laminar region, the heat flux towards the wall is $\dot{q} = -kdT/dy$ and dividing by the expression for the shear stress, $\tau = \mu dc/dy$, yields,

$$\dot{q} = -\tau \frac{k}{\mu} \frac{dT}{dc}.$$

The same relationship is applicable in laminar or turbulent flow if $k/\mu = c_p$ or, expressed slightly differently,

$$\frac{\mu c_p}{k} = \frac{\mu/\rho}{k/\rho c_p} = \frac{v}{\alpha} = 1,$$

where v is the kinematic viscosity, and α is the thermal diffusivity.

The quantity $\mu c_p/k$ is known as the Prandtl number (*Pr*), after the man who first presented the idea of the boundary layer and was one of the pioneers of modern fluid mechanics. For gases, Prandtl numbers are in fact close to unity and for air $Pr = 0.71$ at room temperature. The Prandtl number varies little over a wide range of temperatures: approximately 3 per cent from 300-2000 K. We want a relation between the values at the wall (at which $T = T_\omega$ and $c = 0$) and those in the free stream. To get this, we integrate the expression for dT from the wall to the free stream,

$$-dT = \frac{1}{c_p}\frac{\dot{q}}{\tau}dc,$$

where the relation between heat transfer and shear stress has been taken as the same for both the laminar and the turbulent portions of the boundary layer. The assumption being made is that the mechanisms of heat and momentum transfer are similar. Equation can be integrated from the wall to the freestream (conditions "at ∞"):

$$-\int_\omega^\infty dT = \frac{1}{c_p}\int_\omega^\infty \left(\frac{\dot{q}}{\tau}\right)dc,$$

where $\dot{q}/\tau$ and Cp are assumed constant.

Carrying out the integration yields,

$$T_\omega - T_\infty = \frac{\dot{q}_\omega}{\tau_\omega}\frac{c_\infty}{c_p},$$

where c_∞ is the velocity and c_p is the specific heat. In Equation, $\dot{q}_w$ is the heat flux to the wall and τ_w is the shear stress at the wall. The relation between skin friction (shear stress) at the wall and heat transfer is thus,

$$\frac{\dot{q}_\omega}{\rho_\infty c_p (T_\omega - T_\infty) c_\infty} = \frac{\tau_\omega}{\rho_\infty c_\infty^2}.$$

The quantity,

$$\frac{\tau_w}{\frac{1}{2}\rho_\infty c_\infty^2}$$

is known as the skin friction coefficient and is denoted by *cf*. The skin friction coefficient has been tabulated (or computed) for a large number of situations. If we define a non-dimensional quantity,

$$\frac{\dot{q}_\omega}{\rho_\infty c_p (T_\omega - T_\infty)} = \frac{h(T_\infty - T_\omega)}{\rho_\infty c_p (T_\omega - T_\infty)} = \frac{h}{\rho_\infty c_p c_\infty} = St,$$

known as the Stanton Number, we can write an expression for the heat transfer coefficient, h as,

$$h \approx \rho_\infty c_p c_\infty \frac{cf}{2}.$$

Equation above provides a useful estimate of h, or $\dot{q}_\omega$, based on knowing the skin friction, or drag. The direct relationship between the Stanton Number and the skin friction coefficient is,

$$St = \frac{C_f}{2}.$$

The relation between the heat transfer and the skin friction coefficient,

$$\dot{q}_\omega \approx \frac{\tau_\omega c_p (T_\omega - T_\infty)}{c_\infty}$$

is known as the Reynolds analogy between shear stress and heat transfer. The Reynolds analogy is extremely useful in obtaining a first approximation for heat transfer in situations in which the shear stress is "known."

USE OF THE REYNOLDS ANALOGY

An example of the use of the Reynolds analogy is in analysis of a heat exchanger. One type of heat exchanger has an array of tubes with one fluid flowing inside and another fluid flowing outside, with the objective of transferring heat between them. To begin, we need to examine the flow resistance of a tube. For fully developed flow in a tube, it is more appropriate to use an average velocity $\bar{c}$ and a bulk temperature T_B. Thus, an approximate relation for the heat transfer is,

$$\dot{q}_\omega \approx \tau_\omega c_p \frac{T_B - T_\omega}{\bar{c}}$$

The fluid resistance (drag) is all due to shear forces and is given by $\tau_\omega A_\omega = D$, where A_ω is the tube "wetted" area (perimeter x length). The total heat transfer, $\dot{Q}$, is $\dot{q}_\omega A_\omega$, so that,

$$\frac{\dot{Q}}{P}=\frac{c_p\left(T_B-T_\omega\right)}{\bar{c}^2}$$

The power, *P*, to drive the flow through a resistance is given by the product of the drag and the velocity, $D\bar{c}$, so that,

$$\frac{\dot{Q}}{P}=\frac{c_p\left(T_B-T_\omega\right)}{\bar{c}^2}$$

The mass flow rate is given by $\dot{m}=\rho\bar{c}A$, where *A* is the cross sectional area. For a given mass flow rate and overall heat transfer rate, the power scales as $\bar{c}^2$ or as $1/A^2$, *i.e.*,

$$P\propto\frac{\dot{Q}\dot{m}^2}{\rho^2 c_p(T_B-T_w)}\frac{1}{A^2}.$$

Equations show that to decrease the power dissipated, we need to decrease $\bar{c}^2$, which can be accomplished by increasing the cross-sectional area. Two possible heat exchanger configurations are sketched in Figure; the one on the right will have a lower loss.

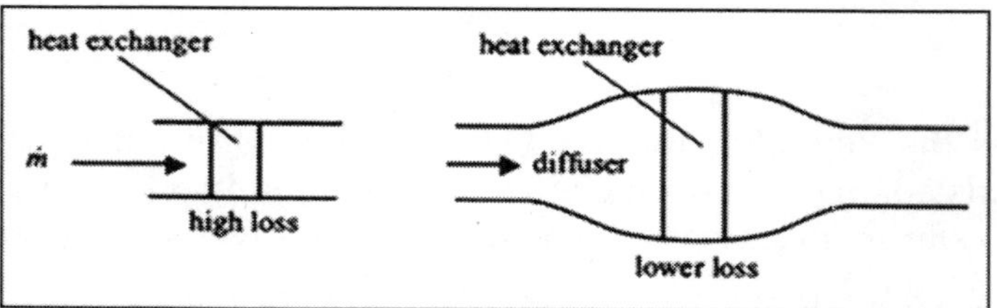

Fig. Heat Exchanger Configurations

To recap, there is an approximate relation between skin friction (momentum flux to the wall) and heat transfer called the Reynolds analogy that provides a useful way to estimate heat transfer rates in situations in which the skin friction is known. The relation is expressed by,

$$St=\frac{C_f}{2},$$

or

$$\frac{\textit{Heat flux to wall}}{\textit{convectd heat flux}}=\frac{\textit{Momentum flux to wall}}{\textit{Convected momentum flux}},$$

or

$$\frac{\dot{q}_W}{\rho_\infty c_\infty c_p\left(T_\infty-T_\omega\right)}=\frac{\tau_\omega}{\rho_\infty c_\infty^2}.$$

The Reynolds analogy can be used to give information about scaling of various effects as well as initial estimates for heat transfer. It is emphasized that it is a useful tool based on a hypothesis about the mechanism of heat transfer and shear stress and not a physical law.

VERTICAL FLATE PLATE IN FREE CONVECTION HEAT TRANSFER

A simple formula to estimate the natural convection heat transfer coefficient on a flat plate was presented in a preceding column.

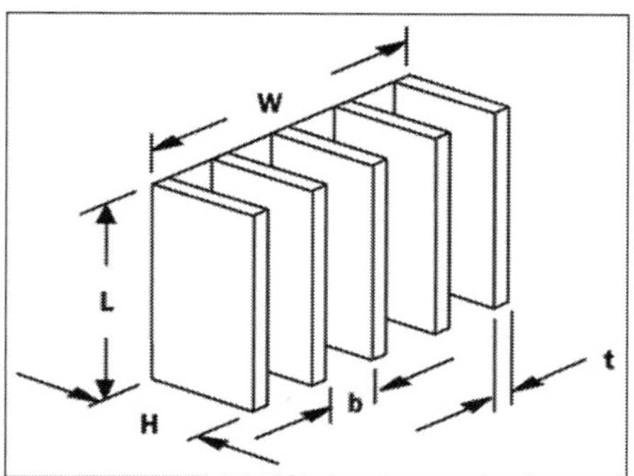

Fig. Vertically Oriented Array of Parallel flat Plates.

If, instead of a single plate, we wish to consider a closely spaced array of vertically oriented parallel flat plates, a different formula is required. The geometry to be considered appears in Figure and is typical of that for a vertically mounted natural convection heat sink.

If the plates are spaced far enough apart, the natural convection boundary layers that develop along the surface of the plates will not interfere with each other. In this case, the heat transfer coefficients would be the same as for a single flat plate. However, if the heated plates are long enough in the natural convection flow direction (L), the boundary layers on adjacent plates will merge. As the plates are moved closer together, the natural convection heat transfer coefficient will fall below the single plate value.

Elenbaas, the first to investigate this behaviour, obtained an experimental correlation to estimate the convection heat transfer coefficient under these conditions. The Elenbaas correlation is of the form

$$Nu = \frac{h.b}{k} = \frac{Ra}{24}\left(1 - e^{-\frac{35}{Ra}}\right)0.75$$

with the Rayleigh number, Ra, given by

$$Ra = \frac{\rho^2 g_c \beta c_p b^4 \Delta T}{\mu k L}$$

where,

b = gap between adjacent plates

c_p= specific heat at constant pressure

g_c= gravitational acceleration

h = heat transfer coefficient

H= height of plate from base

k = thermal conductivity of air

L = length of plate in vertical flow direction

Nu= Nusselt number (dimensionless)

Ra= Rayleigh number (dimensionless)

T_a= temperature of ambient air

T_s= temperature of heated surface

T = temperature difference from surface to air

r= coefficient of thermal expansion for air

p = density of air

= dynamic viscosity of air

e = 2.718, base of natural logarithmTo illustrate the application of these equations we will consider the array of plates (*i.e.*, heat sink configuration) shown in Figure below with the following dimensions (all in mm): L = 100, W = 100, H = 50, t = 2. The value of b, the gap between adjacent fins, may be calculated using,

$$b = \frac{W - N_f.t}{N_f - 1}$$

where W is the width of the base, N_f is the number of plates or fins, and t is the plate or fin thickness. It is also necessary to specify both the ambient air temperature, T_a, entering the array of plates and the surface temperature, Ts, of the plates. The values of air properties to be used are determined at the mean film temperature which is defined as the average of $T_a$and T_s.

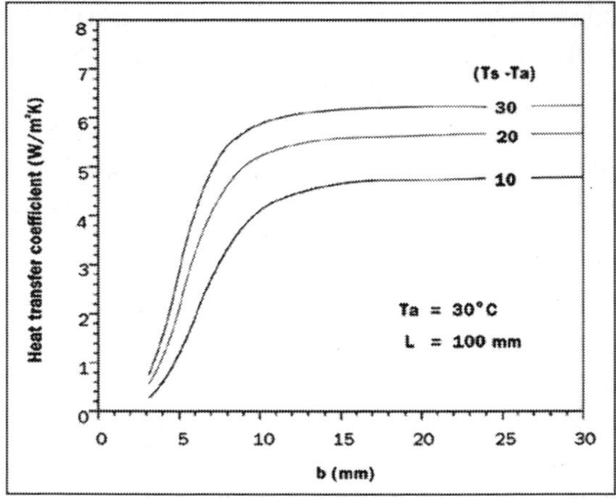

Fig. Natural Convection heat Transfer Coefficients for Array of Plates.

The values of heat transfer coefficient obtained using equation 1 for an ambient air temperature of 30°C and plate surface temperatures of 60, 50, and 40°C appear in Figure. As may be seen, the natural convection heat transfer coefficient increases substantially as the gap between fins increases from 3 to 15 mm, and then flattens out with further increases in gap. It may be noted that the values shown for a gap of 20 mm are within a few per cent of those obtained using a correlation for an individual vertical plate.

Of course, we are interested in the total heat transfer rate from the array of plates. This quantity depends not only on the natural convection heat transfer coefficient but also on the number of plates or fins, Nf. The total heat transfer rate is given by,

$$q = h \cdot A \cdot \Delta T = h \cdot \left(N_f \cdot 2H \cdot L \cdot \eta\right) \cdot \left(T_s - T_a\right)$$

For the sake of simplicity we will use a value of 1 for fin efficiency, η , in equation. Figure shows the values of q normalized with respect to base width, W. As the value of b decreases, the number of fins or plates increases, resulting in a greater total heat transfer area. Since the value of h decreases as the value of b decreases, there will be an optimum value of b which will maximize the product of the total plate area and the heat transfer coefficient, thereby maximizing the total heat transfer rate as seen in Figure.

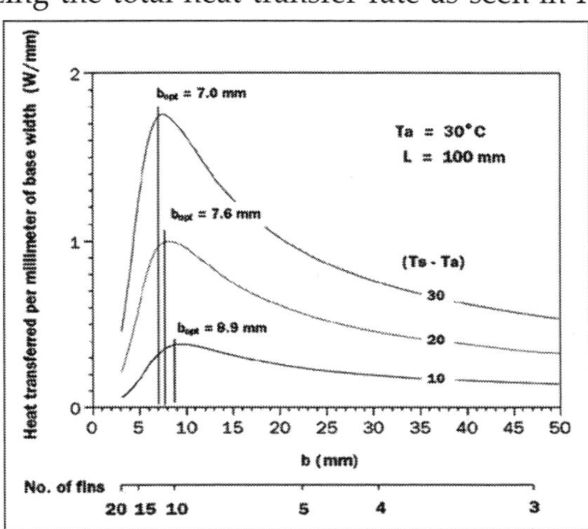

Fig. Total Heat Transfer Rate Per mm of Heat Sink base Width.

Elenbaas established that the optimum spacing, b_{opt}, between the plates in air could be obtained by setting the channel Rayleigh number (based upon plate spacing b and channel aspect ratio b/L) equal to about 50,

$$\frac{b_{opt}}{L}.Gr_{opt} \cdot \Pr = \frac{b}{L} \cdot \frac{\beta g_c \rho^2 b^3 \Delta T}{\mu^2} \cdot \frac{\mu Cp}{k} \cong 50$$

This relationship may then be used to solve for the optimum spacing b_{opt} as given by equation .

$$b_{opt} = 2.659\left(\frac{\mu L k}{\beta g_c \Delta T c_p}\right)0.25\rho - 0.5$$

The vertical lines shown in Figure represent the values of b_{opt} obtained for the example calculations using equation .

Those readers who wish to explore this topic further are referred to reference 2 for an excellent discussion of the subject as well as additional natural convection correlations for a variety of boundary conditions.

6

Mass Transfer

Mass transfer is the net movement of mass from one location, usually meaning a stream, phase, fraction or component, to another. Mass transfer occurs in many processes, such as absorption, evapouration, adsorption, drying, precipitation, membrane filtration, and distillation. Mass transfer is used by different scientific disciplines for different processes and mechanisms. The phrase is commonly used in engineering for physical processes that involve diffusive and convective transport of chemical species within physical systems.

Some common examples of mass transfer processes are the evapouration of water from a pond to the atmosphere, the purification of blood in the kidneys and liver, and the distillation of alcohol. In industrial processes, mass transfer operations include separation of chemical components in distillation columns, absorbers such as scrubbers, adsorbers such as activated carbon beds, and liquid-liquid extraction. Mass transfer is often coupled to additional transport processes, for instance in industrial cooling towers. These towers couple heat transfer to mass transfer by allowing hot water to flow in contact with hotter air and evapourate as it absorbs heat from the air.

Mass transfer finds extensive application in chemical engineering problems. It is used in reaction engineering, separations engineering, heat transfer engineering, and many other sub-disciplines of chemical engineering.

The driving force for mass transfer is typically a difference in chemical potential, when it can be defined, though other thermodynamic gradients may couple to the flow of mass and drive it as well. A chemical species moves from areas of high chemical potential to areas of low chemical potential.

Thus, the maximum theoretical extent of a given mass transfer is typically determined by the point at which the chemical potential is uniform. For single phase-systems, this usually translates to uniform concentration throughout the phase, while for multiphase systems chemical species will often prefer one phase over the others and reach a uniform chemical potential only when most of the chemical species has been absorbed into the preferred phase, as in liquid-liquid extraction.

While thermodynamic equilibrium determines the theoretical extent of a given mass transfer operation, the actual rate of mass transfer will depend on additional factors including the flow patterns within the system and the diffusivities of the species in each phase. This rate can be quantified through the calculation and application of mass transfer coefficients for an overall process.

MIXTURES OF MASS TRANSFER

Mass transfer always involves mixtures. Consequently, we must account for the variation of physical properties which normally exist in a given system. When a system contains three or more components, as many industrial fluid streams do, the problem becomes unwidely very quickly. The conventional engineering approach to problems of multicomponent system is to attempt to reduce them to representative binary (*i.e.,* two component) systems.

In order to understand the future discussions, let us first consider definitions and relations which are often used to explain the role of components within a mixture.

CONCENTRATION OF SPECIES

Concentration of species in multicomponent mixture can be expressed in many ways. For species A, mass concentration denoted by r_A is defined as the mass of A,m_A per unit volume of the mixture.

$$\rho_A = \frac{m_A}{V}$$

The total mass concentration density ρ is the sum of the total mass of the mixture in unit volume:

$$\rho = \sum_i \rho_i$$

where ρ_i is the concentration of species i in the mixture.

Molar concentration of, A, C_A is defined as the number of moles of A present per unit volume of the mixture.

By definition,

$$\text{Numberof moles} = \frac{\text{massof } A}{\text{moleculaweight of } A}$$

$$n_A = \frac{m_A}{M_A}$$

Therefore from above equations:

$$C_A = \frac{n_A}{V} = \frac{\rho_A}{M_A}$$

For ideal gas mixtures,

$$n_A = \frac{p_A V}{RT}$$ [from Ideal gas law PV = nRT]

$$C_A = \frac{n_A}{V} = \frac{p_A}{RT}$$

where p_A is the partial pressure of species A in the mixture. V is the volume of gas, T is the absolute temperature, and R is the universal gas constant.

The total molar concentration or molar density of the mixture is given by:

$$C = \sum_i C_i$$

RELATION AMONG MOLAR FLUXES

For a binary system containing A and B, from Equation (5),

$$N_A = J_A + y_A N$$

or $J_A = N_A + y_A N$

Similarly,

$$J_B = N_B + y_B N$$

Adding Equation gives,

$$J_A + J_B = N_A + N_B - (y_A + y_B) N$$

By definition $N = N_A + N_B$ and $y_A + y_{B-} = 1$.

Therefore equation becomes,

$J_A + J_{B-} = 0$

$J_{A-} = -J_B$

$$CD_{AB} \frac{dy_A}{dz} = -CD_{BA} \frac{dy_B}{dZ}$$

From $y_A + y_{B-} = 1$

$dy_A = -\,dy_B$

Therefore Equation becomes,

$D_{AB} = D_{BA}$

This leads to the conclusion that diffusivity of A in B is equal to diffusivity of B in A.

DIFFUSIVITY

Fick's law proportionality, D_{AB}, is known as mass diffusivity (simply as diffusivity) or as the diffusion coefficient. D_{AB} has the dimension of L^2 / t, identical to the fundamental dimensions of the other transport properties: Kinematic viscosity, $\nu\eta = (\mu/\rho)$ in momentum transfer, and thermal diffusivity, $\alpha\ (= k / \rho C_{\rho})$ in heat transfer.

Diffusivity is normally reported in cm^2 / sec; the SI unit being m^2 / sec.

Diffusivity depends on pressure, temperature, and composition of the system. In table, some values of D_{AB} are given for a few gas, liquid, and solid systems.

Diffusivities of gases at low density are almost composition independent, incease with the temperature and vary inversely with pressure. Liquid and solid diffusivities are strongly concentration dependent and increase with temperature.

General Range of Values of Diffusivity

Gases : 5×10^{-6} ——————— $1 X 10^{-5}$ m^2 / sec.
Liquids : 10^{-6} ——————— 10^{-9} m^2 / sec.
Solids : 5×10^{-14} ——————— $1 X 10^{-10}$ m^2 / sec.

In the absence of experimental data, semitheoretical expressions have been developed which give approximation, sometimes as valid as experimental values, due to the difficulties encountered in experimental measurements.

FICK'S LAW OF DIFFUSION

Diffusion occurs in response to a concentration gradient expressed as the change in concentration due to a change in position, $\partial C/\partial x$. The local rule for movement orflux J is given by Fick's 1st law of diffusion:

$$J = -\chi\frac{\partial C}{\partial x}$$

in which the flux J [cm^{-2} s^{-1}] is proportional to the diffusivity X [cm^2/s] and the negative gradient of concentration, $\partial C/\partial x$ [cm^{-3} cm^{-1}] or [cm^{-4}]. The negative sign indicates that J is positive when movement is down the gradient, *i.e.*, the negative sign cancels the negative gradient along the direction of positive flux.

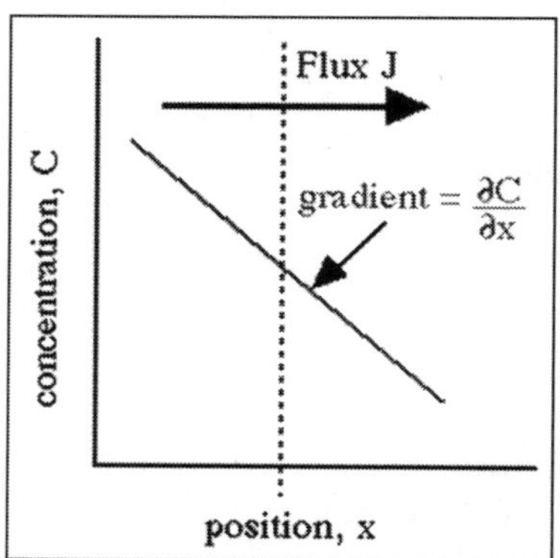

The flux J is driven by the negative gradient $\partial C/\partial x$ in the direction of increasing x.

Example:

Arbitrary units

Consider a local concentration of 10^6 units per cm^3 which drops by 10 per cent over a distance of 1 cm. Then the gradient is -10^5 [units cm^{-4}]. Assume the diffusivity is 10^3 [cm^2/s].

Then the flux J equals:

$J = -(10^3\ [cm^2/s])(-10^5\ [units\ cm^{-4}]) = 10^8\ units\ cm^{-2}\ s^{-1}$

For light, the diffusivity is proportional to the diffusion length D [cm] and the speed of light c:

$$X = cD$$

where $D = 1/(3\ \mu_s(1-g))$. The units of velocity [cm/s] times the units of length [cm] yield the units of diffusivity [cm^2/s]. The following example describes the local diffusion of red light in milk.

Example:

Optical energy

Consider a local fluence rate F of 1 W/cm^2 in milk ($n = 1.33$, $\mu_s(1-g) = 10\ cm^{-1}$).

- The local concentration of optical energy is $C = F/c = (1\ W/cm^2)/(2.2\times10^{10}\ cm/s) = 4.4\times10^{-11}\ J\ cm^{-3}$.
- Assume the concentration drops by 10 per cent over a distance of 3 mm. Then the gradient $\partial C/\partial x$ is $-(0.10)(4.4\times10^{-11}\ cm^{-3})/(0.3\ cm) = -1.5\times10^{-11}\ J\ cm^{-4}$.
- The diffusivity X equals cD where $D = 1/(3(10\ cm^{-1})) = 0.033$ cm. Therefore $X = (2.2\times10^{10}\ cm/s)(0.033\ cm) = 7.5\times10^{10}\ cm^2/s$.
- Then the flux J along the direction x equals $-X\partial C/\partial x = -(7.5\times10^{10}\ cm^2/s)(-1.5\times10^{-11}\ J\ cm^{-4}) = 0.11\ J\ cm^{-2}\ s^{-1} = 0.11\ W\ cm^{-2}$.

For optical diffusion, Fick's 1st law is expressed as the energy flux J [W cm^{-2}] proportional to the diffusion constant D [cm] and the negative fluence gradient dF/dx:

$$J = -D\frac{\partial F}{\partial x}$$

which was obtained by substituting cD for x and substituting F/c for C. The factors c and1/c cancel to yield the above equation.

DEFINITIONS OF QUANTITIES

According to Fick's Law:

J= –D * DC/Dx

Every symbol stands for a different quantity.

J– THE MASS FLUX

Flux is the movement of objects from one point to another in a given time. The flux is what we are measuring when studying diffusion.

Units: the units of moles / (time * area)

Note: area has units of length*length (or $length^2$)

Example: $mol/(h*ft^2)$, $mol/(s*m^2)$

D– DIFFUSIVITY

The diffusivity is the constant that describes how fast or slow an object diffuses.

Units: the units of area / time
Example: ft^2/h, or cm^2/s

Δ - DELTA

Delta is the Greek letter that represents the words "change in." When the symbol D is used, it refers to subtracting the beginning value from the final value of the symbol following the Δ.

- *For example*: $\Delta C = C_2 - C_1$ = the last C – the first C, where C represents concentration.

C – CONCENTRATION

Concentration is the amount of mass in a given volume.

The symbol Δ C refers to the change in concentration from when the object had not diffused at all, to the final concentration when the object was done diffusing.

Units: the units of amount of substance/volume

Note: volume is the representation of size in three dimensions. Therefore it has the units of length*length*length (or $length^3$)

Example: mol / cm^3, mol/L

X – DISTANCE

This refers to the distance that the object is diffusing.

The symbol Δx refers to the distance between where the object started and where it ended up after it diffused.

Units: the units of length
Example: m, cm, ft
Therefore, returning to the equation,
J = –D * DC/Dx

It was described that the flux was equal to the negative diffusivity times the change in concentration divided by the change in distance.

7

Fundamentals of Thermodynamics

INTRODUCTION

Thermodynamics is a science and an engineering tool, which is used to describe processes that involve changes in temperature, transformation of energy, and relationships between heat and work. It can be regarded as a generalization of an enormous body of empirical evidence. It is extremely general that there are no hypotheses made which is concerning the structure and type of matter that we deal with. It is used to describe the performance of propulsion systems, power generation systems, refrigerators, fluid flow, combustion and many other phenomena. The focus of thermodynamics in engineering is on the production of work, in the form of kinetic energy or shaft power from different sources of heat. For the most part the heat will be the result of combustion processes, but this is not always the case. The course content can be viewed in terms of a "propulsion chain" as shown in Figure.

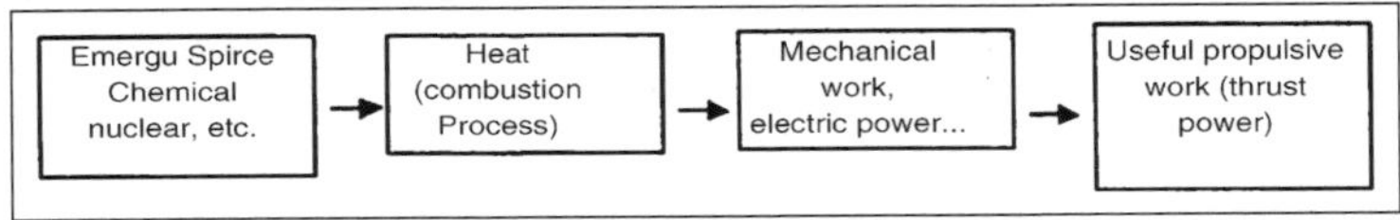

Figure

In the above figure, we see a progression from an energy source to useful propulsive work. In terms of the different blocks Parts I and II are mainly about how to progress from the second block to the third, We will start with the progression from heat to work, examples of which are given in Figure.

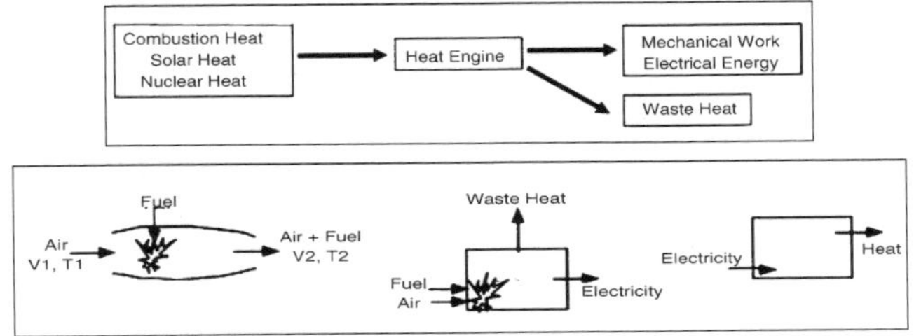

Figure

PRINCIPLE OF THERMODYNAMICS

In science, thermodynamics is concerned with the mathematical modelling of the real world. In order that the mathematical deductions are consistent, we need some precise definitions of the basic concepts.

THE CONTINUUM MODEL

Matter may be described at a molecular or microscopic level using the techniques of statistical mechanics and kinetic theory. For engineering purposes, however, we want "averaged" information, *i.e.*, a macroscopic, not a microscopic, description. There are two reasons for this. First, a microscopic description of an engineering device may produce too much information to manage.

For example, 1mm^3 of air at standard temperature and pressure contains 10^{16} molecules each of which has a position and a velocity. Typical engineering applications involve more than 10^{20} molecules. Second, and more importantly, microscopic positions and velocities are generally not useful for determining how macroscopic systems will act or react unless, for instance, their total effect is integrated. We therefore neglect the fact that real substances are composed of discrete molecules and model matter from the start as a smoothed-out continuum.

The information we have about a continuum represents the microscopic information averaged over a volume. Classical thermodynamics is concerned only with continua.

THE CONCEPT OF A "SYSTEM"

A thermodynamic system is a quantity of matter of fixed identity, around which we can draw a boundary. The boundaries may be fixed or moveable. Work or heat can be transferred across the system boundary. Everything outside the boundary is the surroundings. When working with devices such as engines it is often useful to define the system to be an identifiable volume with *flow* in and out. This is termed a control volume. An example is shown in Figure. A closed system is a special class of system with boundaries that matter cannot cross. Hence the principle of the conservation of mass is automatically satisfied whenever we employ a closed system analysis. This type of system is sometimes termed a control mass.

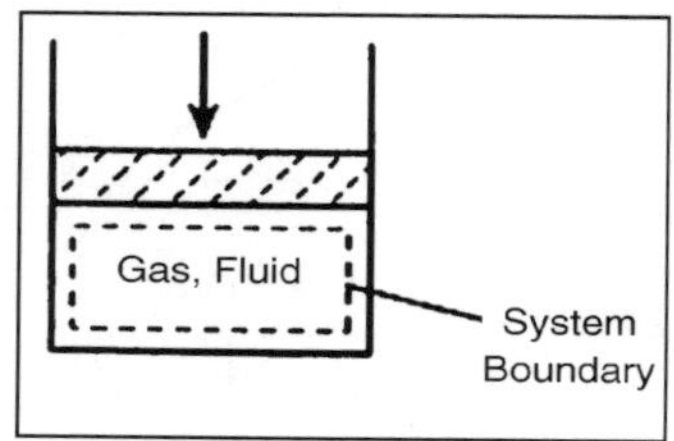

Fig. Piston (Boundary) and Gas (System)

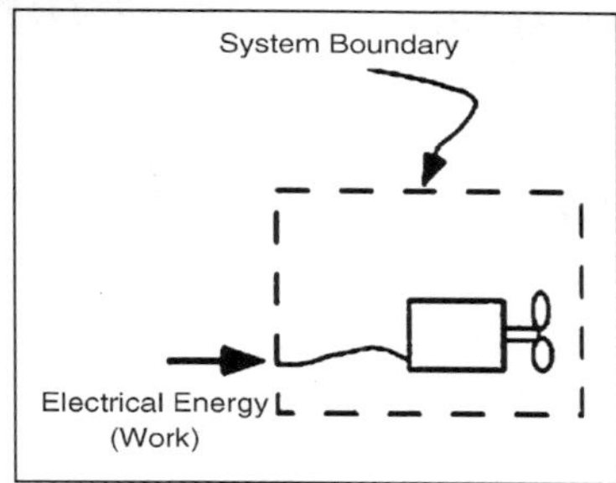

Fig. Boundary Around Electric Motor (System)

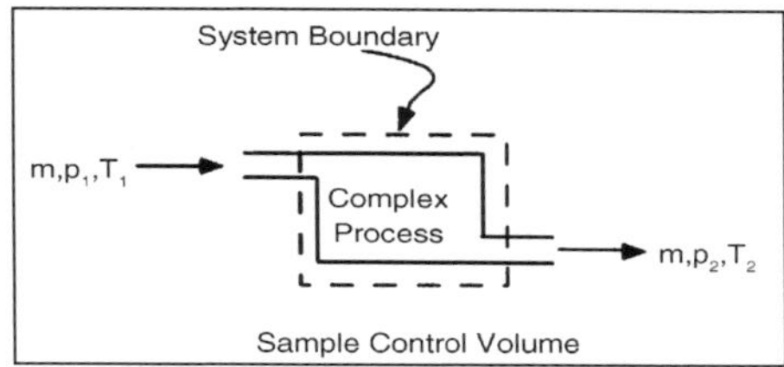

Fig. Sample Control Volume

THE CONCEPT OF A "STATE"

The thermodynamic state of a system is defined by specifying values of a set of measurable properties sufficient to determine all other properties. For fluid systems, typical properties are pressure, volume and temperature. More complex systems may require the specification of more unusual properties. As an example, the state of an electric battery requires the specification of the amount of electric charge it contains.

Properties may be extensive or intensive. Extensive properties are additive. Thus, if the system is divided into a number of sub-systems, the value of the property for the whole system is equal to the sum of the values for the parts. *Volume* is an extensive property. Intensive properties do not depend on the quantity of matter present. *Temperature* and *pressure* are intensive properties. Specific properties are extensive properties per unit mass and are denoted by lower case letters.

E.g.:

$$\text{Specific Volume} = V/m = n$$

Specific properties are intensive because they do not depend on the mass of the system. The properties of a simple system are uniform throughout. In general, however, the properties of a system can vary from point to point. We can usually analyse a general system by sub-dividing it (either conceptually or in practice) into a number of simple systems in each of which the properties are assumed to be uniform. It is important to note that properties describe states *only* when the system is in equilibrium.

THE CONCEPT OF "EQUILIBRIUM"

The state of a system in which properties have definite, unchanged values as long as external conditions are unchanged is called an equilibrium state.

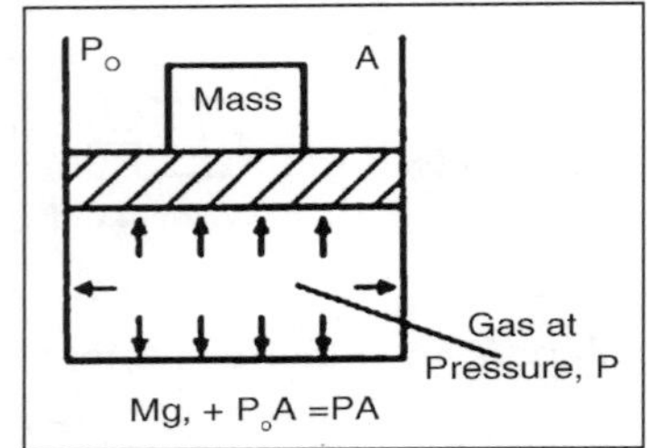

Fig. Mechanical Equilibrium

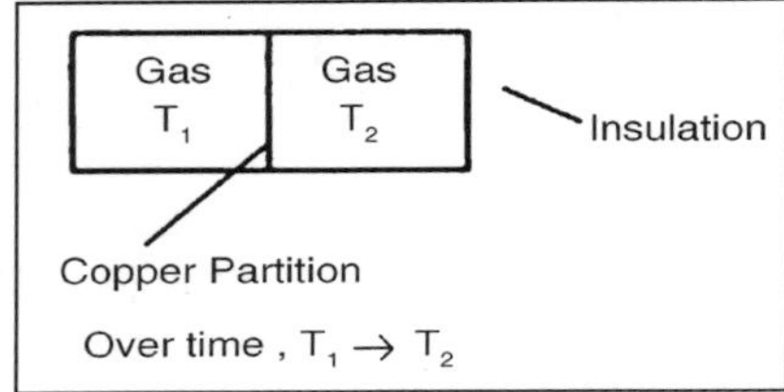

Fig. Thermal Equilibrium

A system in thermodynamic equilibrium satisfies:

- Mechanical equilibrium (no unbalanced forces)
- Thermal equilibrium (no temperature differences)
- Chemical equilibrium.

THE CONCEPT OF A "PROCESS"

If the state of a system changes, then it is undergoing a process. The succession of states through which the system passes defines the path of the process. If at the end of the process, the properties have returned to their original values, the system has undergone a cyclic process or a cycle. We note that even if a system has returned to its original state and completed a cycle, the state of the surroundings may have changed.

Quasi-Equilibrium Processes

We are often interested in charting thermodynamic processes between states on thermodynamic coordinates. If a process involves finite, unbalanced forces, the system can pass through non-equilibrium states, which we cannot treat. An extremely useful idealization, however, is that only "infinitesimal" unbalanced forces exist, so that the process can be viewed as taking place in a series of "quasi-equilibrium" states. For this to be true the process must be slow in relation to the time needed for the system to come to equilibrium internally. For a gas at conditions of interest to us, a given molecule can undergo roughly 10^{10} molecular collisions per second, so that, if ten collisions are needed to come to equilibrium, the equilibration time is on the order of 10^{-9} seconds. This is generally much shorter than the time scales associated with the bulk properties of the flow (say the time needed for a fluid particle to move some significant fraction of the length of the device of interest). Over a large range of parameters, therefore, it is a very good approximation to view the thermodynamic processes as consisting of such a succession of equilibrium states, which we can chart. The figures, below demonstrate the use of thermodynamics coordinates to plot isolines, lines along which a property is constant. They include constant temperature lines, or isotherms, on a $p - v$ diagram, constant volume lines, or isochors on a $T - p$ diagram, and constant pressure lines, or isobars, on a $T - v$ diagram for an ideal gas.

Real substances may have phase changes (water to water vapour, or water to ice, for example), which we can also plot on thermodynamic coordinates.

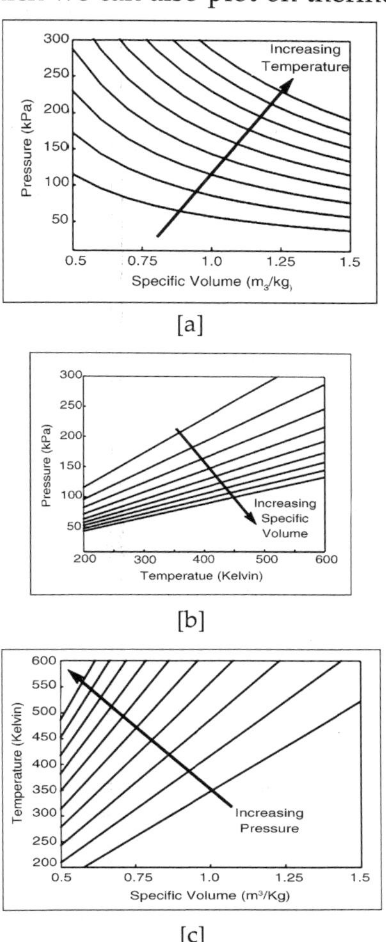

Fig. Thermodynamics Coordinates and Isolines for an Ideal Gas

EQUATIONS OF STATE

It is an experimental fact that two properties are needed to define the state of any pure substance in equilibrium or undergoing a steady or quasi-steady process. Thus for a simple compressible gas like air,

$$P = P(v, T), \quad \text{or} \quad v = v(P, T), \quad \text{or} \quad T = T(P, v),$$

where v is the volume per unit mass, I/p. In words, if we know v and T we know P, etc.

Any of these is equivalent to an equation $f(P, v, T) = 0$ which is known as an equation of state. The equation of state for an ideal gas, which is a very good approximation to real gases at conditions that are typically of interest for aerospace applications is

$$P\overline{v} = \mathcal{R}T,$$

where $\overline{v}$ is the volume per mol of gas and $\mathcal{R}$ is the "Universal Gas Constant 8.31 KJ/Kmol-K.

A form of this equation which is more useful in fluid flow problems is obtained if we divide by the molecular weight, $\mathcal{M}$:

$$Pv = RT, \quad \text{or} \quad P = \rho RT$$

where R is R/M, which has a different value for different gases due to the different molecular weights. For air at room conditions R = 0.287 KJ/Kg – K

SCOPE

This interdisciplinary scope of thermodynamics provides us a forum for presenting new ideas in continuum and quasi-continuum modelling of systems with a large number of degrees of freedom and sufficient complexity to require thermodynamic closure. Major emphasis is placed on papers attempting to bridge the gap between discrete and continuum approaches as well as micro- and macro-scales, by means of homogenization, statistical averaging and other mathematical tools aimed at the judicial elimination of small time and length scales. The scope of thermodynamics is particularly in contributions on focusing of a simultaneous description of complex systems at several disparate scales. Papers presenting and explaining new experimental findings are highly encouraged. The scope welcomes numerical studies aimed at understanding the physical nature of the phenomena.

Potential subjects range from boiling and turbulence to plasticity and earthquakes. Studies of fluids and solids with nonlinear and non-local interactions, multiple fields and multi-scale responses, nontrivial dissipative properties and complex dynamics are expected to have a strong presence. An incomplete list of featured topics includes: active solids and liquids, nano-scale effects and molecular structure of materials, singularities in fluid and solid mechanics, polymers, elastomers and liquid crystals, rheology, cavitation and fracture, hysteresis and friction, mechanics of solid and liquid phase transformations, composite, porous and granular media, scaling in statics and dynamics, large scale processes and geomechanics, stochastic aspects of mechanics. This would also attract papers addressing the very foundations of thermodynamics and kinetics of continuum processes. Of special interest are contributions to the emerging areas of biophysics and biomechanics of cells, bones and tissues leading to new continuum and thermodynamical models.

PROCESS OF THERMODYNAMIC

When the system undergoes change from one thermodynamic state to final state due change in properties like temperature, pressure, volume etc., the system is said to have undergone thermodynamic process. Various types of thermodynamic processes are: isothermal process, adiabatic process, isochoric process, isobaric process and reversible process. These have been described below:

ISOTHERMAL PROCESS

Isothermal process is a change or process that occurs at a constant temperature. During an isothermal process, energy enters or leaves the system to maintain a constant temperature. The collapse of a protostar down the Hayashi track, and the collapse of a star at the end of its life to become a white dwarf, is examples of isothermal changes.

When the system undergoes change from one state to the other, but its temperature remains constant, the system is said to have undergone isothermal process. For instance, in our example of hot water in thermos flask, if we remove certain quantity of water from the flask, but keep its temperature constant at 50 degree Celsius, the process is said to be isothermal process.

Another example of isothermal process is latent heat of vaporization of water. When we heat water to 100 degree Celsius, it will not start boiling instantly. It will keep on absorbing heat at constant temperature; this heat is called latent heat of vaporization. Only after absorbing this heat water at constant temperature, water will get converted into steam.

A process which occurs at constant temperature is called isothermal. For an ideal gas constant T means constant $U(T)CvT$ and thus from the first law we have

$$0 = Q_{A \to B} - W_{A \to B}$$

$$W_{A \to B} = Q_{A \to B}$$

This is the most efficient way of generating work from heat. However in order to keep T constant a finite heat transfer requires changes in p and V according to the equation of state. Specifically adding heat isothermally will require the system to do positive work and hence that the volume increases while the pressure decreases. keeping the product constant. We shall see that we cannot conceive of an engine which simply performs an isothermal expansion at all times because its volume would have to grow without bounds. Thus other processes must be included in the thermal cycle of an engine and these eventually limit the efficiency of the conversion of heat into mechanical work. We can easily evaluate the work performed by an ideal gas in an isothermal process by using

$$W_{A \to B} = \int_A^B pdV = nRT \int_A^B \frac{dV}{V} = nRT \ln \frac{V_B}{V_A}$$

and this same expression obviously also holds for the heat transfer because of above equations

$$Q_{A \to B} = \int_A^B pdV = nRT \int_A^B \frac{dV}{V} = nRT \quad \frac{V_B}{V_A}$$

ADIABATIC PROCESS

The process, during which the heat content of the system or certain quantity of the matter remains constant, is called as adiabatic process. Thus in adiabatic process no transfer of heat between the system and its surroundings takes place. The wall of the system which does not allows the flow of heat through it, is called as adiabatic wall, while the wall which allows the flow of heat is called as diathermic wall.

An adiabatic process is one in which no heat is gained or lost by the system. The first law of thermodynamics with Q = 0 shows that all the change in internal energy is in the form of work done. This puts a constraint on the heat engine process leading to the adiabatic condition shown below. This condition can be used to derive the expression for the work done during an adiabatic process.

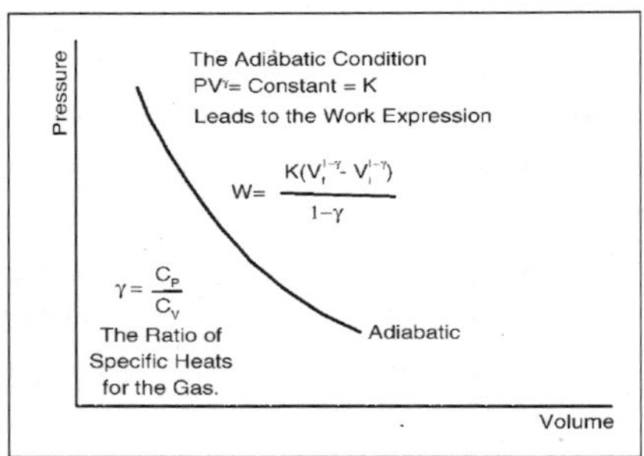

Figure

The ratio of the specific heats $\gamma = C_P/C_V$ is a factor in determining the speed of sound in a gas and other adiabatic processes as well as this application to heat engines. This ratio $\gamma = 1.66$ for an ideal monoatomic gas and $\gamma = 1.4$ for air, which is predominantly a diatomic gas as shown in figure.

DEVELOPMENT OF THE ADIABATIC CONDITION

The difference from the Ideal gas law

$P\Delta V + V\Delta P \quad = nR\Delta T$

Since Q = 0 for the adiabatic case, the first law of thermodynamics becomes

$\Delta U \qquad = 2P\Delta V$

and the expression for specific heat

$$C_V = \frac{1}{n}\frac{\Delta U}{\Delta T} \text{ leads to}$$

$$n\Delta T = \frac{-P}{C_V}\Delta V$$

Using $R = C_P - C_V$, the gas law gives

$$n\Delta T = \frac{P\Delta V + V\Delta P}{C_P - C_V} = \frac{-P}{C_V}\Delta V$$

With rearrangement this becomes $\dfrac{\Delta P}{P} + \dfrac{C_P}{C_V}\dfrac{\Delta V}{V} = 0$

The limit $\dfrac{dP}{P} + \gamma\dfrac{dV}{V} = 0$ can be integrated to give

In $(P) + \gamma \ln (V)$ = constant

$$\gamma = \frac{C_P}{C_V}$$

Using log combination rules, this can be rearranged to $\ln(PV^\gamma)$ = constant and therefore PV^γ = constant which may be called the adiabatic condition.

Since the adiabatic constant γ for a gas is the ratio of the specific heats as indicated above, it depends upon the effective number of degrees of freedom in the molecular motion. It can in fact be expressed as $\gamma = (f + 2)/f$ where f is the number of degrees of freedom in the molecular motion. For a monatomic gas like helium, f = 3 and $\gamma = 5/3$. For diatomic molecules like N_2 and O_2, you include two degrees of rotational freedom, so f = 5 and $\gamma = 1.4$. Since almost all of the atmosphere is nitrogen and oxygen, $\gamma = 1.4$ can be used for air in such calculations as the speed of sound. For polyatomic molecules, there are three degrees of rotational freedom and this along with the three translational degrees of freedom would give $\gamma = 4/3$. But in the specific heats of polyatomic molecules, there is evidence of contribution by vibrational degrees of freedom, and this would further reduce the value of γ.

ISOCHORIC PROCESS

The process, during which the volume of the system remains constant, is called as isochoric process. Heating of gas in a closed cylinder is an example of isochoric process. An isochoric process, also called an isometric process or an isovolumetric process, is a process during which volume remains constant. The name is derived from the Greek *isos,* "equal", and *khora,*"place."

If an ideal gas is used in an isochoric process, and the quantity of gas stays constant, then the increase in energy is proportional to an increase in temperature and pressure. Take for example a gas heated in a rigid container: the pressure and temperature of the gas will increase, but the volume will remain the same.

In the ideal Otto cycle we found an example of an isochoric process when we assume an instantaneous burning of the gasoline-air mixture in an internal combustion engine car. There is an increase in the temperature and the pressure of the gas inside the piston while the volume remains the same.

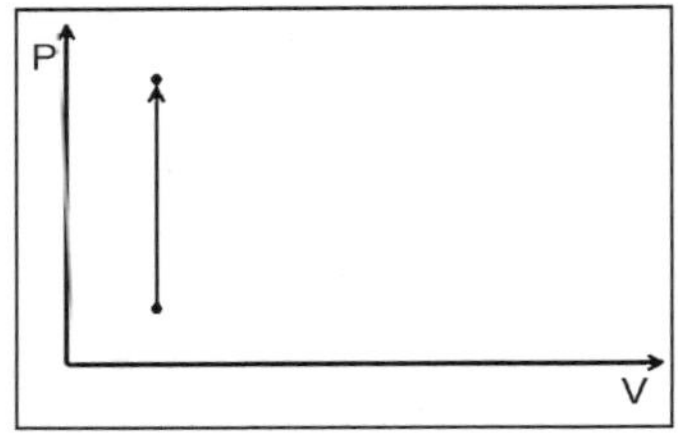

Figure

Pressure Volume Diagram: In this diagram, pressure increases, but volume remains constant.

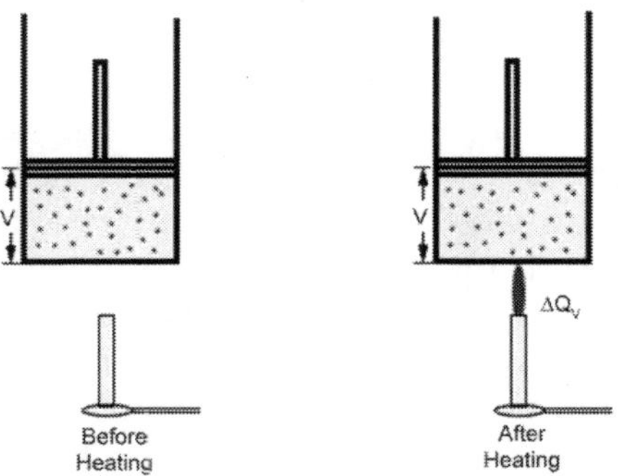

Figure

In the above figure.

Consider a cylinder fitted with a frictionless piston. An ideal gas is enclosed in the cylinder. The piston is fixed at a particular position so that the volume of cylinder remains constant during the supply of heat.

Let DQ amount of heat is added to the system. Addition of heat causes the following changes in the system:

Internal energy increases from U_1 to U_2.

Volume of the system remains unchanged.
Temperature increases from T_1 K to T_2 K.
Pressure increases from P_1 to P_2.
No work is performed.

If the volume stays constant ($\Delta V = 0$), this implies that the process does no pressure-volume work, since such work is defined by

$$\Delta W = P\Delta V$$

where P is pressure (no minus sign; this is work done *by* the system).

By applying the first law of thermodynamics, we can deduce that ΔU the change in the system's internal energy, is

$$\Delta U = Q$$

for an isochoric process: all the heat being transferred to the system is added to the system's internal energy, U. If the quantity of gas stays constant, then this increase in energy is proportional to an increase in temperature,

$$Q = n\,C_V \Delta T$$

where C_V is molar specific heat for constant volume.

On a pressure volume diagram, an isochoric process appears as a straight vertical line. Its thermodynamic conjugate, an isobaric process would appear as a straight horizontal line.

ISOBARIC PROCESS

The process during which the pressure of the system remains constant is called as isobaric process. Example: Suppose there is a fuel in piston and cylinder arrangement. When this fuel is burnt the pressure of the gases is generated inside the engine and as more fuel burns more pressure is created. But if the gases are allowed to expand by allowing the piston to move outside, the pressure of the system can be kept constant.

The constant pressure and volume processes are very important. The Otto and diesel cycle, which are used in the petrol and diesel engine respectively, have constant volume and constant pressure processes. In practical situations ideal constant pressure and constant pressure processes cannot be achieved.

An isobaric process is a thermodynamic process in which the pressure stays constant; $\Delta P = 0$. The heat transferred to the system does work but also changes the internal energy of the system:

$$Q = \Delta E + W \quad (1)$$

according to the first law of thermodynamics, where W is work done *by* the system, E is internal energy, and Q is heat. Pressure-volume work (*by* the system) is defined as:

$$W = \Delta(PV) \quad (2)$$

but since pressure is constant, this means that

Applying the ideal gas law, equation (2) becomes

$$W = nR\Delta T \quad (3)$$

assuming that the quantity of gas stays constant (*e.g.* no phase change during a chemical reaction).

Since it is generally true that

$$\Delta E = nC_V\Delta T \qquad (4)$$

then substituting equations (3) and (4) into equation (1) produces:

$$Q = nC_V\Delta T + nR\Delta T$$
$$= n(C_V + R)\Delta T$$

The quantity in parenthesis is equivalent to the molar specific heat for constant pressure:

$$C_P = C_V + R$$

and if the gas involved in the isobaric process is monatomic then

$$C_V = (3/2)R \text{ and}$$
$$C_P = (5/2)R.$$

An isobaric process is shown on a P-V diagram as a straight horizontal line, connecting the initial and final thermostatic states. If the process moves towards the right, then it is an expansion. If the process moves towards the left, then it is a compression.

ISENTROPIC PROCESS

In thermodynamics, an isentropic process is one during which the entropy of the system remains constant, isentropic process is usually one that is both internally reversible (can be reversed without leaving a trace on its surroundings) and adiabatic (no heat transfer occurs between the system and its surroundings).

Most devices (pumps, turbines, etc.) are essentially adiabatic, and all perform best when irreversibilities are minimized. Therefore, isentropic processes are often used to model real processes. Also, because an isentropic process represents the most efficient process possible, they are used to define efficiencies of real processes.

For an ideal gas, pressure, density, and temperature at two states of an isentropic process are related by the equations:

p2/p1 = (ρ2/ρ1)^γ = (T2/T1)^(γ/(γ-1))

where p is pressure, ρ is density, T is temperature, γ is the specific heat ratio (1.4 for air), and 1 and 2 represent different points in the process. These relations are especially useful in aerodynamics, where most areas of flow around an airfoil (beyond the boundary layer) can be assumed to be isentropic.

From the second law of thermodynamics, an isentropic process of liquids and solids is also an isothermal (constant temperature) process.

Isentropic conditions also occur at stagnation points - a fluid condition defined by the point at which a flow is isentropically brought to zero velocity.

During an isentropic process, the state of the thermodynamic variables of a gas can change. If the specific heat capacity is a constant value, the gas is said to be calorically perfect and if the specific heat capacity changes, the gas is said to becalorically imperfect. At subsonic and low supersonic Mach numbers, air is calorically perfect. But under low hypersonic conditions, air is calorically imperfect. Derived flow variables, like the speed of sound and the isentropic flow relations are slightly different for a calorically imperfect gas than the conditions predicted for a calorically perfect gas because some of the energy of the flow excites the vibrational modes of the diatomic molecules of nitrogen and oxygen in the air.

Isentropic flows occur when the change in flow variables is small and gradual, such as the ideal flow through the nozzle shown above. The generation of sound waves is an isentropic process. A supersonic flow that is turned while the flow area increases is also isentropic. We call this an isentropic expansion because of the area increase. If a supersonic flow is turned abruptly and the flow area decreases, shock waves are generated and the flow is irreversible. The isentropic relations are no longer valid and the flow is governed by the oblique or normal shock relations.

THERMODYNAMIC SYSTEMS

An important concept in thermodynamics is the "system". Everything in the universe except the system is known as surroundings. A system is the region of the universe under study. A system is separated from the remainder of the universe by a boundary which may be imaginary or not, but which by convention delimits a finite volume. The possible exchanges of work, heat, or matter between the system and the surroundings take place across this boundary. Boundaries are of four types: fixed, moveable, real, and imaginary.

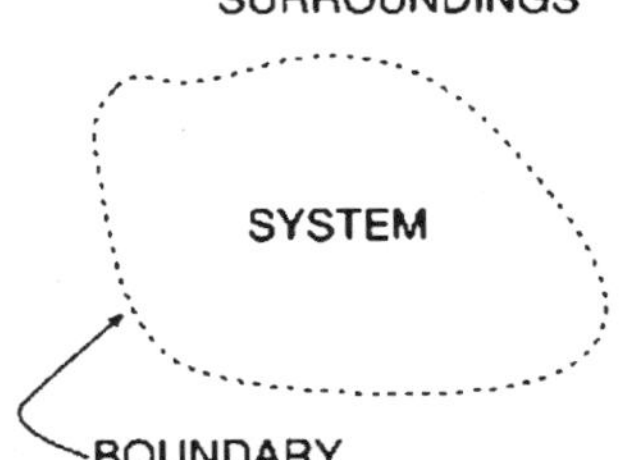

Fig. Thermodynamic Systems

Basically, the "boundary" is simply an imaginary dotted line drawn around a volume of something when there is going to be a change in the internal energy of that something. Anything that passes across the boundary that effects a change in the internal energy of the something needs to be accounted for in the energy balance equation. That something can be the volumetric region surrounding a single atom resonating energy, such as Max Planck defined in 1900; it can be a body of steam or air in a steam engine, such as Sadi Carnot defined in 1824; it can be the body of a tropical cyclone, such as Kerry Emanuel theorized in 1986 in the field of atmospheric thermodynamics; it could also be just one nuclide (i.e. a system of quarks) as some are theorizing presently in quantum thermodynamics.

For an engine, a fixed boundary means the piston is locked at its position; as such, a constant volume process occurs. In that same engine, a moveable boundary allows the piston to move in and out. For closed systems, boundaries are real while for open system boundaries are often imaginary. There are five dominant classes of systems:

- *Isolated Systems*–matter and energy may not cross the boundary
- *Adiabatic Systems*–heat must not cross the boundary
- *Diathermic Systems*–heat may cross boundary
- *Closed Systems*–matter may not cross the boundary
- *Open Systems*–heat, work, and matter may cross the boundary

As time passes in an isolated system, internal differences in the system tend to even out and pressures and temperatures tend to equalize, as do density differences. A system in which all equalizing processes have gone practically to completion, is considered to be in a state of thermodynamic equilibrium.

In thermodynamic equilibrium, a system's properties are, by definition, unchanging in time. Systems in equilibrium are much simpler and easier to understand than systems which are not in equilibrium. Often, when analysing a thermodynamic process, it can be assumed that each intermediate state in the process is at equilibrium. This will also considerably simplify the situation. Thermodynamic processes which develop so slowly as to allow each intermediate step to be an equilibrium state are said to be reversible processes.

Thermodynamic Parameters

The central concept of thermodynamics is that of energy, the ability to do work. As stipulated by the first law, the total energy of the system and its surroundings is conserved. It may be transferred into a body by heating, compression, or addition of matter, and extracted from a body either by cooling, expansion, or extraction of matter.

For comparison, in mechanics, energy transfer results from a force which causes displacement, the product of the two being the amount of energy transferred. In a similar way, thermodynamic systems can be thought of as transferring energy as the result of a generalized force causing a generalized displacement, with the product of the two being the amount of energy transferred. These thermodynamic force-displacement pairs are known as conjugate variables. The most common conjugate thermodynamic variables are pressure-volume (mechanical parameters), temperature-entropy (thermal parameters), and chemical potential-particle number (material parameters).

Thermodynamic Instruments

There are two types of thermodynamic instruments, the meter and the reservoir. A thermodynamic meter is any device which measures any parameter of a thermodynamic system. In some cases, the thermodynamic parameter is actually defined in terms of an idealized measuring instrument. For example, the zeroth law states that if two bodies are in thermal equilibrium with a third body, they are also in thermal equilibrium with each other. This principle, as noted by James Maxwell in 1872, asserts that it is possible to measure temperature. An idealized thermometer is a sample of an ideal gas at constant pressure.

From the ideal gas law PV=nRT, the volume of such a sample can be used as an indicator of temperature; in this manner it defines temperature. Although pressure is defined mechanically, a pressure-measuring device, called a barometer may also be constructed from a sample of an ideal gas held at a constant temperature. A calorimeter is a device which is used to measure and define the internal energy of a system.

A thermodynamic reservoir is a system which is so large that it does not appreciably alter its state parameters when brought into contact with the test system. It is used to impose a particular value of a state parameter upon the system. For example, a pressure reservoir is a system at a particular pressure, which imposes that pressure upon any test system that it is mechanically connected to. The earth's atmosphere is often used as a pressure reservoir.

It is important that these two types of instruments are distinct. A meter does not perform its task accurately if it behaves like a reservoir of the state variable it is trying to measure. If, for example, a thermometer were to act as a temperature reservoir it would alter the temperature of the system being measured, and the reading would be incorrect. Ideal meters have no effect on the state variables of the system they are measuring.

Thermodynamic States

When a system is at equilibrium under a given set of conditions, it is said to be in a definite state. The state of the system can be described by a number of intensive variables and extensive variables. The properties of the system can be described by an equation of state which specifies the relationship between these variables. State may be thought of as the instantaneous quantitative description of a system with a set number of variables held constant

Thermodynamics is an experimental science based on a small number of principles that are generalizations made from experience. It is concerned only with macroscopic or large-scale properties of matter and it makes no hypotheses about the small-scale or microscopic structure of matter. From the principles of thermodynamics one can derive general relations between such quantities as coefficients of expansion, compressibilities, specific heat capacities, heats of transformation, and magnetic and dielectric coefficients, especially as these are affected by temperature. The principles of thermodynamics also tell us which of these relations must be determined experimentally in order to completely specify all the properties of the system...

Thermodynamics is complementary to kinetic theory and statistical thermodynamics. Thermodynamics provides relationships between physical properties of any system once certain measurements are made. Kinetic theory and statistical thermodynamics enable one to calculate the mangitudes of these properties for those systems whose energy states can be determined.

THERMODYNAMIC EQUILIBRIUM

A thermodynamic system is said to be in thermodynamic equilibrium when it is in thermal equilibrium, mechanical equilibrium, and chemical equilibrium. The local state of a system at thermodynamic equilibrium is determined by the values of its intensive parameters, *e.g.*, temperature and pressure.

Thermodynamic equilibrium is characterized by the minimum of a thermodynamic potential, such as the Helmholtz energy (for systems at constant temperature and volume):

$$F = U - T \cdot S$$

Or as the Gibbs energy (for systems at constant pressure and volume):

$$G = H - T \cdot S$$

A system that is in equilibrium experiences no changes when it is isolated from its surroundings.

MECHANICAL EQUILIBRIUM

We are now ready to consider objects in equilibrium. There are two conditions to equilibrium for most objects.

The first condition is stated by Newton's first law:

$$\sum \vec{F} = 0$$

The second condition of equilibrium is

$$\sum \vec{\tau} = 0$$

Basically, we say that a body is in equilibrium if the vector sum of the forces and torques are zero.

Usually, we will need to look at both translational equilibrium and rotational equilibrium. The rule of thumb is that if the body is a point source, or all of the forces act at the same point on the body, then we do not need to consider rotational equilibrium. If the forces act at different points on the body, then we must take rotation into account.

Centre of Mass

In order to simplify our calculations, it is convenient to talk in terms of the centre of mass, or centre of gravity, of a body.

We define the centre of mass to be the weighed average of the components of the body

$$\vec{X}_{CM} = \frac{\sum m_i \vec{x}_i}{\sum m_i} = \frac{\sum m_i x_i}{M} \quad (1)$$

If the object is a continuous mass distribution, we replace the summation with an integral. Also, notice that the centre of mass is a vector. The usefulness of the centre of mass is that we will often need to calculate the torque of the centre of mass about some point. It can be shown that in terms of the centre of mass, the equilibrium conditions reduce to

$$\sum \vec{F}_{ext} = 0$$

$$\sum \vec{\tau}_{ext} = \sum \vec{R} \times \vec{F}_{ext} = 0 \qquad (2)$$

where F_{ext} is all of the forces external to the body, and R is the distance from the centre of mass.

Example

Locate the centre of mass of the machine part in the diagram below

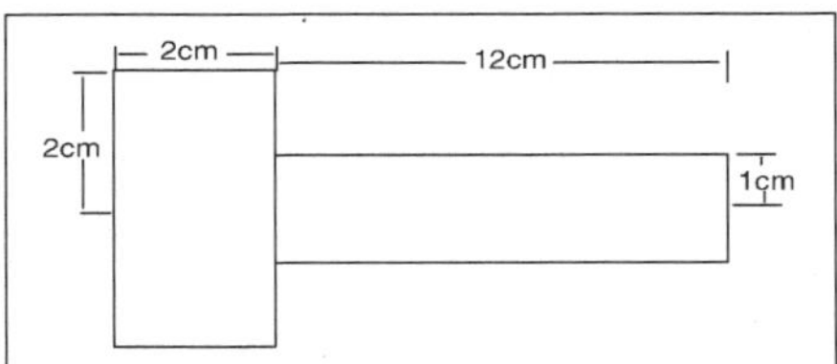

By symmetry, the centre of mass lies along the axis and the centre of mass of each part is midway between its ends. The volume of the disk is 8p cm^3 and that of the rod is 12p cm^3. Since the weights of the two parts are proportional to their volumes,

$$\frac{W_{disk}}{W_{rod}} = \frac{8\pi}{12\pi} = \frac{2}{3}$$

Taking the origin to be at the left face of the disc on the axis, we have

$$x_1 = 1 \text{ cm } x_2 = 8 \text{ cm}$$

and so

$$X_{CM} = \frac{W_{disk} X_1 + W_{rod} X_2}{W_{disk} + W_{rod}}$$

$$= \frac{\frac{2}{3} W_{rod}(1 \text{ cm}) + W_{rod}(8 \text{ cm})}{\frac{2}{3} W_{rod} + W_{rod}}$$

$$= 5.2 \text{ cm}$$

ENGINEERING THERMODYNAMICS

DEFINITION

Engineering Thermodynamics is the study of energies and energy transfer in the systems. It consists of methods and constructions that are used to "account" for macroscopic energy transfer. In fact energy accounting is an appropriate synonym for thermodynamics. In much the same way that accountants balance money in and money out of a bank account, rocket scientists simply balance the energy in and out of a rocket engine. Of course just as a bank account's balance is obfuscated by arcane devices such as interest rates and currency exchange, so too is thermodynamics clouded with seemingly difficult concepts such as irreversibility and enthalpy. But, also just like accounting, a careful review of the rules suggests a coherent strategy for maintaining tabs on a particular account.

If a statement about the simplicity of thermodynamics failed to convert would-be students, they may be captured with a few words on the importance of understanding energy transfer in our society. Up until about 150 years ago or so, the earth's economy was primarily fuelled by carbohydrates. That is to say, humans got stuff done by converting food, through a biological process; to fuel we could spend to do work. This was a hindrance to getting things accomplished because, as it turned out, most individuals had to use the brunt of that energy to grow and cultivate more carbohydrates.

Today, we have the luxury, primarily through an understanding of energy, to concentrate our energy production into efficient low maintenance operations. Massive power plants transfer energy to power tools for raising barns. Extremely efficient rocket engines tame and direct massive amounts of energy to blast TV satellites into orbit. This improvement in energy mastery frees humanity's time to engage in more worthwhile activities such as watching cable TV. Understanding energy transfer and energy systems is the next step to destroying the limits to what humanity can next accomplish. The first step is commanding an interest in doing so from an inclined portion of the population.

APPLICATIONS OF ENGINEERING THERMODYNAMICS

All materials can exist in three phases as *solid, liquid,* and *gas*. All one component systems share certain characteristics, so that a study of a typical one component system will be quite useful.

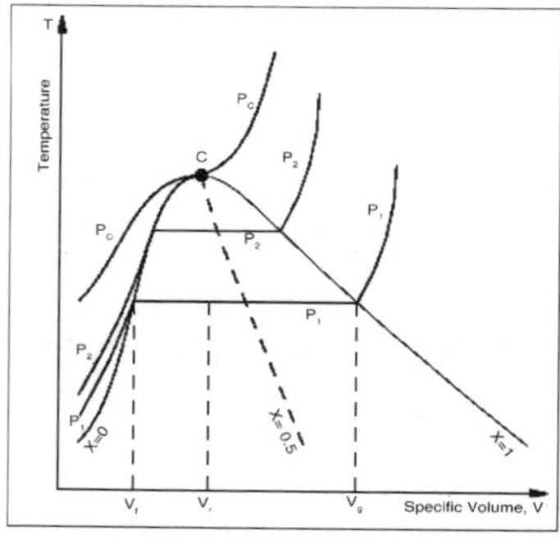

Figure

For this consider the above figure, where the heat is transferred to the substance at constant pressure. In this there are temperature and volume curves with all three pressures constants. In this the three line-curves labelled as p_1, p_2, and p are known as isobars which shows conditions at constant pressure. When the liquid and vapour coexist, it is called a saturated state. There is no change in temperature or pressure when liquid and vapour are in equilibrium, so that the temperature is called *saturation temperature* and the pressure is called *saturation pressure.*

Saturated states are represented by the horizontal lines in the chart. In the temperature range where both liquid and vapour of a pure substance can coexist in equilibrium, for every value of saturated temperature, there is only one corresponding value of saturation pressure. If the temperature of the liquid is lower than the saturation temperature, it is called *subcooled liquid.* If the temperature of the vapour or gas is greater than the saturation temperature it is called *superheated vapour.*

The amount of liquid and vapour in a saturated mixture is specified by its *quality x*, which is the fraction of vapour in the mixture. Thus, the horizontal line representing the vaporization of the fluid has a quality of $x = 0$ at the left endpoint where it is pure liquid and a quality of $x = 1$ at the right endpoint where it is pure vapour. The blue curve in the preceding diagram shows saturation temperatures for saturated liquid *i.e.* where $x = 0$. The green curve in the diagram shows saturation temperatures for saturated vapour *i.e.* where $x = 1$ which are not isobars.

$$x = \frac{m_g}{m_f + m_g}$$

$$v_x = (1 - x)v_f + xv_g$$

$$v_{fg} = v_g - v_f$$

We can say that the point where the solid, liquid and vapour state exist in equilibrium is called as the *triple point*. We further see that as the saturation temperatures increase, the liquid and vapour specific volumes approach each other until the blue and green curves come together and meet at point C on the p_c isobar. At that point C, called the *critical point,* the liquid and vapour states merge together and all their thermodynamic properties become the same. The critical point has a certain temperature T_c, and pressure p_c, which depend on the substance in question. At temperatures above the critical point, the substance is considered a super-heated gas.

Gibbs Phase Rule

Gibbs phase rule states that for a heterogeneous system in equilibrium with *C* components in *P* phases, the degree of freedom is given by:

$$F = C - P + 2.$$

Thus, for a single component system with two phases, there is only one degree of freedom *i.e.*: F = 1 – 2 + 2 F = 1 That is, if you are given either the pressure or temperature of wet steam, you can obtain all the properties, while for superheated steam, which has just one phase, you will need both the pressure and the temperature.

Psychrometry

Psychrometry is the study of air and water vapour mixtures which are used for air conditioning. For this the air is taken as a mixture of nitrogen and oxygen with other gases being small enough so that they can be approximated by more of nitrogen and oxygen without much error.

In psychrometry, the vapour refers to water vapour. For air at normal (atmospheric) pressure, the saturation pressure of vapour is very low. Also, air is far away from its critical point in those conditions. Thus, the air vapour mixture behaves as an ideal gas mixture. If the partial pressure of the vapour is smaller than the saturation pressure for water for that temperature, the mixture is called unsaturated. The amount of moisture in the air vapour mixture is quantified by its *humidity*.

The absolute humidity ω is the ratio of masses of the vapour and air, *i.e.*, $\omega = m_v/m_a$. Now, applying ideal gas equation, $pV = mRT$ for water vapour and for air, we have, since the volume and temperature are the same, $\omega = 0.622\, p_v/p_a$. The ratio of specific gas constants (R in preceding equation) of water vapour to air equals 0.622.

The *relative humidity* φ is the ratio of the vapour pressure to the saturation vapour pressure at that temperature,

i.e., $\varphi = p_v/p_{v,sat}$.

The *saturation ratio* is the ratio of the absolute humidity to the absolute humidity at saturation, or, $\psi = \omega/\omega_{sat}$. It is easy to see that the saturation ratio is very close to the value of relative humidity.

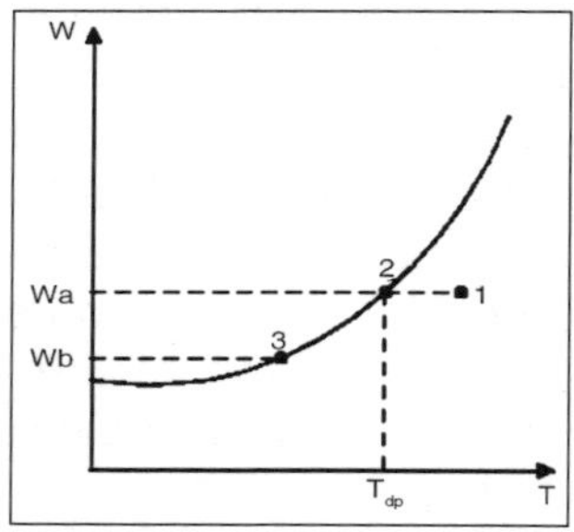

Figure

The above figure shows the value of absolute humidity versus the temperature. The initial state of the mixture is 1, and it is cooled isobarically, and at constant absolute humidity. When it reaches 2, it is saturated, and its absolute humidity is ω_a. Further cooling cause's condensation and the system moves to point 3, where its absolute humidity is ω_b. The temperature at 2 is called the *dew point*.

It is customary to state all quantities in psychrometry per unit mass of dry air. Thus, the amount of air condensed in the above chart when moving from 2 to 3 is $\omega_b - \omega_a$.

Adiabatic Saturation

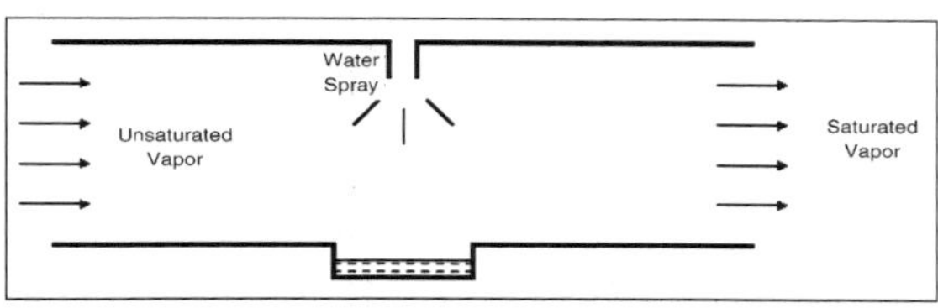

Figure

Consider an unsaturated mixture entering a chamber as shown in figure. Suppose water was sprayed into the stream, so that the humidity increases and it leaves as a saturated mixture. This is accompanied by a loss of temperature due to heat being removed from the air which is used for vaporization. If the water supplied is at the temperature of exit of the stream, then there is no heat transfer from the water to the mixture. The final temperature of the mixture is called *adiabatic saturation temperature.*

Wet Bulb Temperature

The relative humidity of air vapour mixtures is measured by using dry and wet bulb thermometers. The dry bulb thermometer is an ordinary thermometer, while the wet bulb thermometer has its bulb covered by a moist wick. When the mixture flows past the two thermometers, the dry bulb thermometer shows the temperature of the stream, while water evaporates from the wick and its temperature falls. This temperature is very close to the adiabatic saturation temperature if we neglect the heat transfer due to convection.

Psychrometric Chart

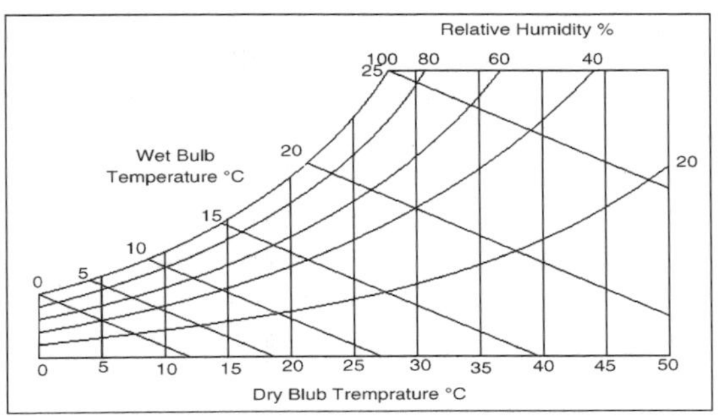

This chart gives the value of absolute humidity versus temperature, along with the enthalpy. From this chart you can determine the relative humidity given the dry and wet bulb temperatures. We have, from the first law, that for a flow system with no heat transfer, the enthalpy is a constant. Now, for the adiabatic saturation process, there is no heat transfer taking place, so that the adiabatic saturation lines are the same as the wet bulb temperature and the constant enthalpy lines.

Air Conditioning

The human body can work efficiently only in a narrow range of conditions. Further, it rejects about 60 W of heat continuously into the surroundings, and more during heavy exercise. The temperature of the body is maintained by the evaporation of sweat from the body. Thus, for comfort, both the temperature and the relative humidity should be low.

Conventional air conditioning consists of setting the humidity at an acceptable level, while reducing the temperature. Reducing the humidity to zero is not the ideal objective. For instance, low humidity leads to issues like high chances of static electricity building up, leading to damage of sensitive electronic equipment. A humidity level of 50 per cent is more acceptable in this case.

The most common method of reducing humidity is to cool the air using a conventional air conditioner working on a reversed Carnot cycle. The vapour that condenses is removed. Now, the air that is produced is very cold, and needs to be heated back up to room temperature before it is released back to the air conditioned area.

Thermodynamic Cycles

The Otto Cycle

The Otto cycle is an idealization of a set of processes used by spark ignition internal combustion engines (2-stroke or 4-stroke cycles).

These engines a) ingest a mixture of fuel and air, b) compress it, c) cause it to react, thus effectively adding heat through converting chemical energy into thermal energy, d) expand the combustion products, and then e) eject the combustion products and replace them with a new charge of fuel and air.

Representation of heat engine as a thermodynamic cycle.

(Ingest mixture of fuel and air)

1′ - 2 Compress mixture quasi-statically and adiabatically

2 - 3 Ignite and burn mixture at constant volume (heat is added)

3 - 4 Expand mixture quasi-statically and adiabatically

4 - 1″ Cool mixture at constant volume (then repeat)

work ~ Force* distance ~ $\int pdv$

fuel use ~ heat added ~ $T_3 - T_2$
efficiency ~ work out/fuel use

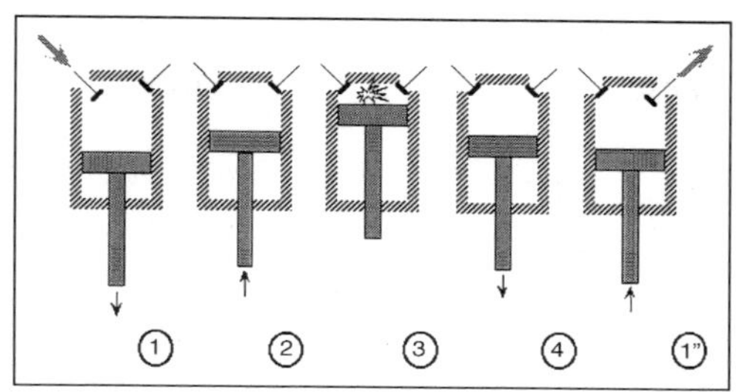

Figure

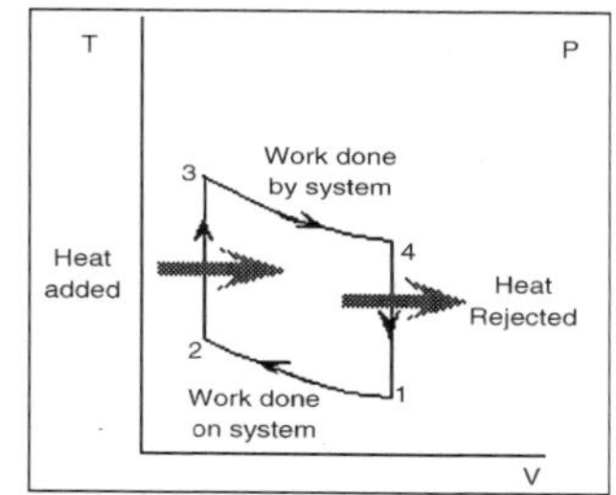

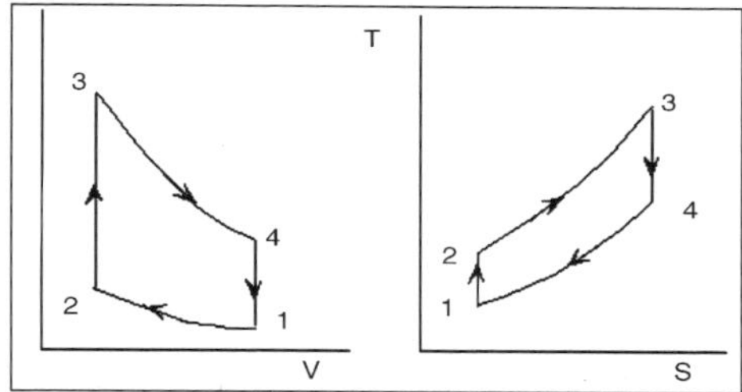

Figure

Method for estimating thermal efficiency & work output (First Law of Thermodynamics)

Net work done by system = work of expansion + work of compression (–) Both expansion and compression are adiabatic so

$$Dw = (u_3 - u_4) - (u_2 - u_1)$$

Assuming an ideal gas with constant c_v

$$Dw = c_v[(T_3 - T_4) - (T_2 - T_1)]$$

While the above expression is accurate, it is not all that useful. We would like to put the expression in terms of the typical design parameters: the compression ratio ($v_1/v_2 = v_4/v_3$), and the heat added during combustion ($q_{comb.} = c_v(T_3 - T_2)$).

For a quasi-static, adiabatic process,

$$\frac{T_2}{T_1} = \left(\frac{v_1}{v_2}\right)^{\gamma-1} = \frac{T_3}{T_4}$$

so we can write the net work as

$$\Delta \mathrm{w} = c_v T_1 \left(\frac{T_4}{T_1} - 1 \right) \left(r^{\gamma-1} - 1 \right)$$

we also know that

$$\frac{T_4}{T_1} = \left(\frac{4}{3} \right) \left(\frac{3}{2} \right) \left(\frac{2}{1} \right) = \frac{3}{2} \text{ since } \frac{2}{1} = \frac{3}{4}$$

so

$$\frac{T_4}{T_1} - 1 = \frac{T_3}{T_2} - 1 = \left(\frac{T_3 - T_2}{T_1} \right) \left(\frac{1}{r^{\gamma-1}} \right)$$

and finally, the result in terms of typical design parameters are:

$$\Delta \mathrm{w} = q_{\text{comb.}} \frac{\left(r^{\gamma-1} - 1 \right)}{r^{\gamma-1}}$$

The thermal efficiency of the cycle is

$$\eta = \frac{\text{network}}{\text{heat input}} = \frac{\Delta w}{q_{\text{comb.}}}$$

$$\eta_{\text{Otto}} = \frac{\left(r^{\gamma-1} - 1 \right)}{r^{\gamma-1}}$$

so

(Rewrite this expression as

$$\eta_{\text{Otto}} = 1 - \frac{T_1}{T_2}$$

showing that the efficiency of an Otto cycle depends only on the temperature ratio of the compression process.).

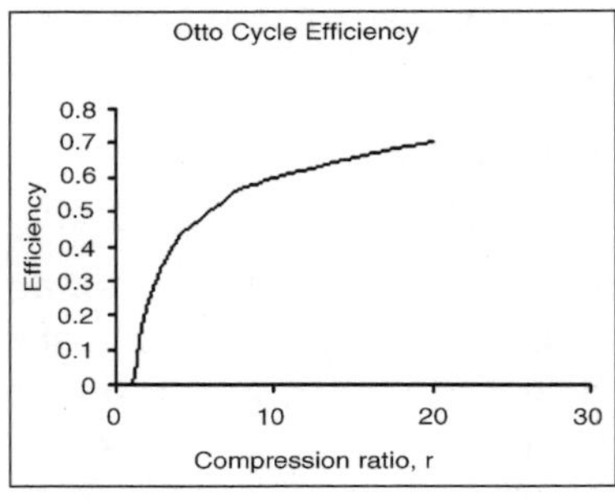

Figure

Brayton Cycle

Brayton cycle is an ideally a set of thermodynamic processes, which is used in turbine engines, for jet propulsion or for generation of electrical power. Components of a gas turbine engine are shown in figure.

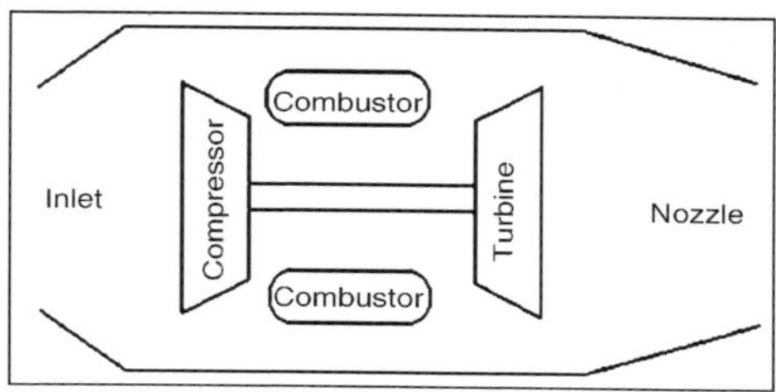

Figure

Schematics of typical gas turbine engine: J57 turbojet with afterburning is shown in figure.

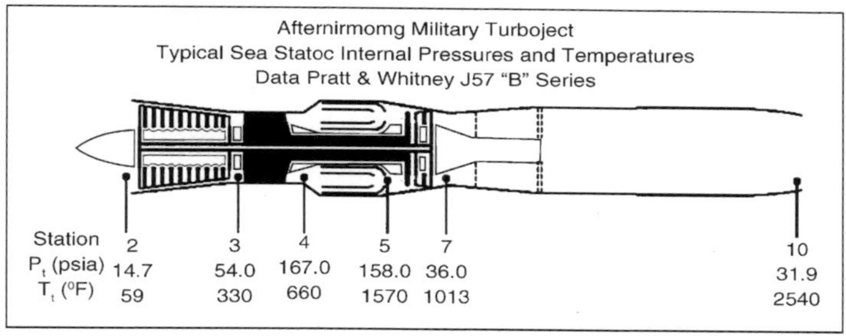

Figure

Process of Thermodynamic cycle

The cycle consists of four processes:

- Quasi-static adiabatic compression in the inlet and compressor
- Constant pressure heat addition in the combustor
- Quasi-static adiabatic expansion in the turbine and exhaust nozzle, and finally
- Constant pressure cooling to get the working fluid back to the initial condition.

1-2 Adiabatic, quasi-static compression in inlet and compressor
2-3 Combust fuel at constant pressure (*i.e.* add heat)
3-4 Adiabatic, quasi-static expansion in turbine

- Take work out and use it to drive the compressor
- Take remaining work out and use it to accelerate fluid for jet propulsion, or to turn a generator for electrical power generation.

4-1 Cool the air at constant pressure
Then repeat

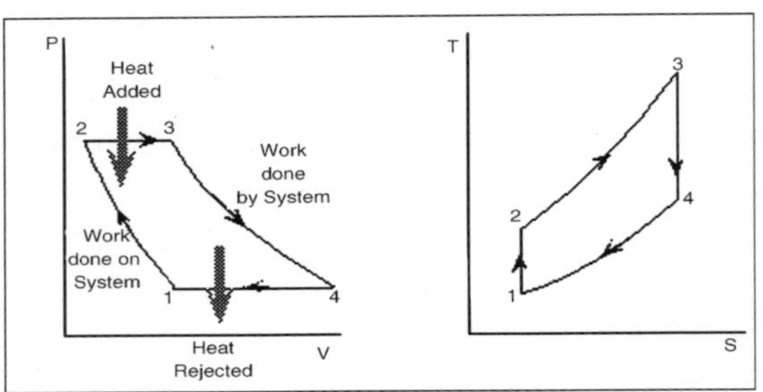

Figure

Estimating the Performance of the Engine

First to derive expressions for the net work and the thermal efficiency of the cycle, and then to manipulate these expressions to put them in terms of typical design parameters so that they will be more useful.

First, remember from the First Law we can show that for any cyclic process heat and work transfers are numerically equal

$Du = q - w$

$u_{\text{final}} = u_{\text{initial}}$ therefore $Du = 0$ and

$q = w$ or $\oint \delta q = \oint \delta w$

This fact is often useful for solving thermodynamic cycles. For instance in this example, we would like to find the net work of the cycle and we could calculate this by taking the difference of the work done all the way around the cycle. Or, since Dq = Dw, we could just as well consider the difference between the heat added to the cycle in process 2-3, and the heat rejected by the cycle in process 4-1.

First Law in terms of enthalpy for an ideal gas undergoing a quasi-static process:

$dq = dh - vdp$ or $dq = c_p dT - vdp$

at constant pressure

$q_{\text{added}} = c_p\, DT$ or $q_{\text{added}} = c_p\,(T_3 - T_2)$

Hat rejected between 4-1:

Similarly

$q_{\text{added}} = c_p\, DT$ or $q_{\text{rejected}} = c_p\,(T_4 - T_1)$

Work done and thermal efficiency:

$Dw = Dq = q_{\text{added}} - q_{\text{rejected}}$

$= c_p[(T_3 - T_2) - (T_4 - T_1)]$

$h_{\text{Brayton}} = (q_{\text{added}} - q_{\text{rejected}})/q_{\text{added}}$

$= [(T_3 - T_2) - (T_4 - T_1)]/(T_3 - T_2)$

Again, while these expressions are accurate, they are not all that useful. We need to manipulate them to put them in terms of typical design parameters for gas turbine engines. For gas turbine engines the most useful design parameters to use for these equations are often the inlet temperature (T_1), the compressor pressure ratio (p_2/p_1), and the maximum cycle temperature, the turbine inlet temperature (T_3).

Rewriting equations in terms of design parameters:

$$\Delta w = c_p T_1 \left(\frac{T_3}{T_1} - \frac{T_2}{T_1} - \frac{T_4}{T_1} + 1 \right)$$

$$\frac{P_2}{P_1} = \frac{P_3}{P_4} = \left(\frac{T_2}{T_1} \right)^{\gamma/\gamma-1} = \left(\frac{T_3}{T_4} \right)^{\gamma/\gamma-1} \quad \text{therefore}$$

$$\frac{T_2}{T_1} = \frac{T_3}{T_4} \text{ and } \frac{T_4}{T_1} = \frac{T_3}{T_2}$$

so

$$\Delta w = C_P T_1 \left[\frac{T_3}{T_1} - \left(\frac{P_2}{P_1} \right)^{\gamma-1/\gamma} - \frac{T_3}{T_2} + 1 \right]$$

with

$$T_2 = T_1 \left(\frac{P_2}{P_1} \right)^{\gamma-1/\gamma}$$

becomes

$$\Delta w = C_P T_1 \left[\frac{T_3}{T_1} - \left(\frac{P_2}{P_1} \right)^{\gamma-1/\gamma} - \frac{T_3}{T_1} \left(\frac{P_2}{P_1} \right)^{-(\gamma-1)/\gamma} + 1 \right]$$

and for the efficiency

$$\eta = 1 - \left(\frac{T_4 - T_1}{T_3 - T_2} \right) = 1 - \left[\frac{\frac{T_4}{T_1} - 1}{\frac{T_3}{T_2} - 1} \right]$$

but from above

$$\frac{T_4}{T_1} = \frac{T_3}{T_2}$$

so

$$h_{Brayton} = 1 - \frac{T_1}{T_2} = 1 - \frac{1}{\left(\frac{P_2}{P_1} \right)^{\gamma-1/\gamma}}$$

Performance plots of Brayton Cycle

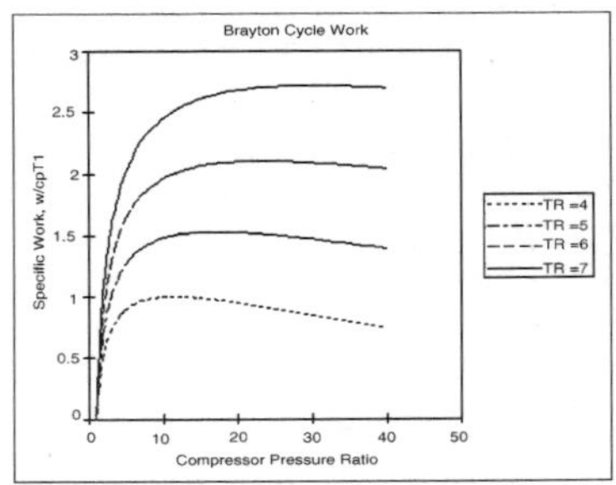

Figure

In the plot above, TR = T_3/T_1. For a given turbine inlet temperature, T_3, there is a compressor pressure ratio that maximizes the work.

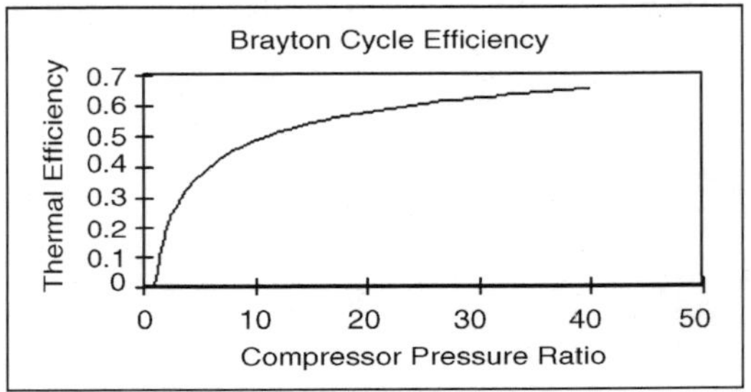

Figure

Generalized Representation of Thermodynamic Cycles

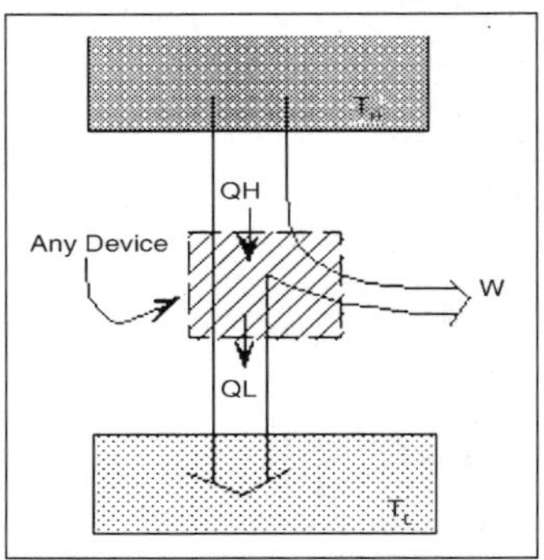

Figure

We see that the heat engines can be represented generally as a transfer of heat from a high temperature reservoir to a device + a rejection of heat from a device to a low temperature reservoir + net work done on surroundings.

Refrigeration Cycles

Refrigeration cycles take in work from the surroundings and transfer heat from a low temperature reservoir to a high temperature reservoir.

Schematically, they look like the diagram given above, but with the direction of the arrows reversed. They can also be recognized on thermodynamic diagrams as closed loops with a counter-clockwise direction of travel. A detailed physical description is given in figure.

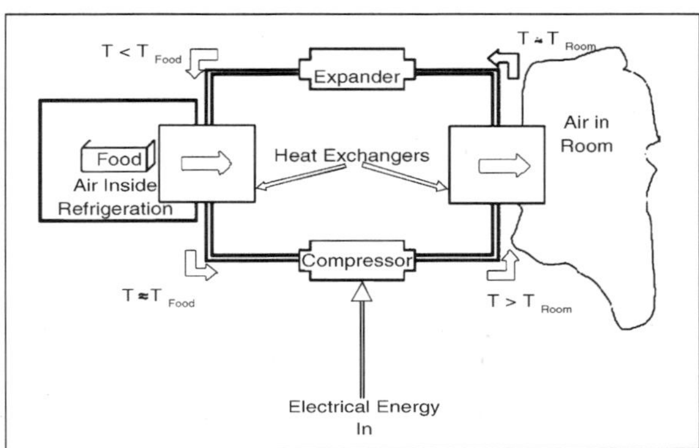

Figure

The objective of a refrigerator is to lower the internal energy of a body at low temperature and transfer that energy to the higher temperature surroundings. It requires work to do this. The medium for the energy exchange is a working fluid that circulates in a loop through a series of devices. These devices act to add and remove energy from the working fluid. Typically the working fluid in the loop is considered the thermodynamic system. Sometimes the fluid used alternates between gas-phase and liquid-phase, but this detail is not important for understanding the basic process. A simplified diagram is shown in figure. As the fluid circulates around the loop, its internal energy is alternately raised and lowered by a series of devices. In this manner, the working fluid is colder than the refrigerator air at one point and hotter than the air in the room at another point. Thus heat will flow in the appropriate direction as shown by the two arrows in the heat exchangers.

Starting in the upper right hand corner of the diagram, first the internal energy is lowered either by passing through a small turbine or through an expansion valve. In these devices, work is done by the refrigerant so its internal energy is lowered. The internal energy is lowered to a point where the temperature of the refrigerant is lower than that of the air in the refrigerator. A heat exchanger is used to transfer energy from the air (and food) in the refrigerator to the cold refrigerant (energy transferred by virtue of a temperature difference only = heat). This lowers the internal energy of the air/food and raises the internal energy of the refrigerant. Then a pump or compressor is used to do work on the refrigerant adding additional energy to it and thus further raising its internal energy. Electrical energy is used to drive the pump or compressor. The internal energy of the refrigerant is raised to a point where its temperature is hotter than the temperature of the room.

The refrigerant is then passed through a heat exchanger (the coils at the back of the refrigerator) so that energy is transferred from the refrigerant to the surroundings. As a result, the internal energy of the refrigerant is reduced and the internal energy of the surroundings is increased. It is at this point where the internal energy of the food and the energy used to drive the compressor or pump are transferred to the surroundings. The refrigerant then continues on to the turbine, repeating the cycle.

THERMODYNAMICS CYCLE

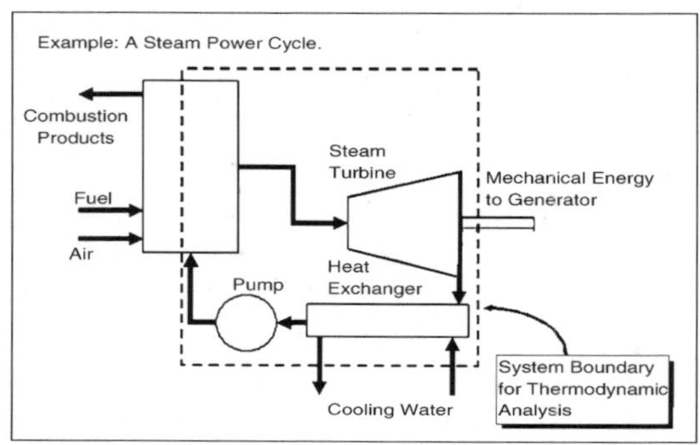

SYSTEMS

A thermodynamic system is that part of the universe that is under consideration. A real or imaginary boundary separates the system from the rest of the universe, which is referred to as the environment. A useful classification of thermodynamic systems is based on the nature of the boundary and the flows of matter, energy and entropy through it.

ISOLATED SYSTEM

It does not exchange heat, matter or work with the environment. An example of an isolated system would be an insulated container, such as an insulated gas cylinder. In reality, a system can never be absolutely isolated from its environment, because there is always at least some slight coupling, even if only via minimal gravitational attraction.

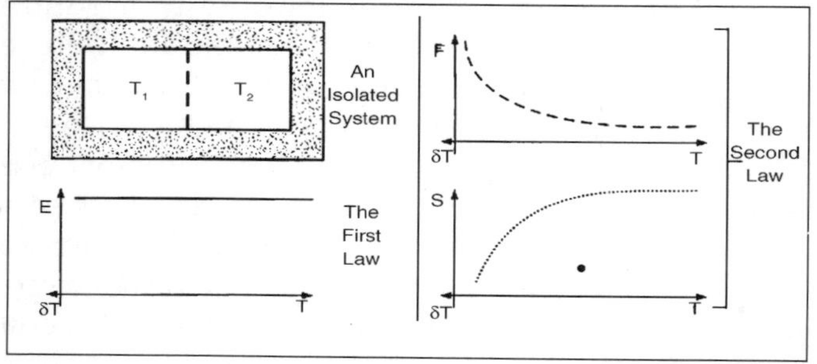

Figure

Figure shows the essence of classical thermodynamics: In a system isolated from the outside world, heat within a gas of temperature, T_2, will flow in time, t, towards a gas of temperature, T_1, where $T_2 > T_1$ and $T = T_2 - T_1$, thus conserving the system's total energy, E (via the first law of thermodynamics), all the while its free energy F, decreases, its entropy, S, rises (via the second law of thermodynamics), and T 0 at equilibrium.

CLOSED SYSTEM

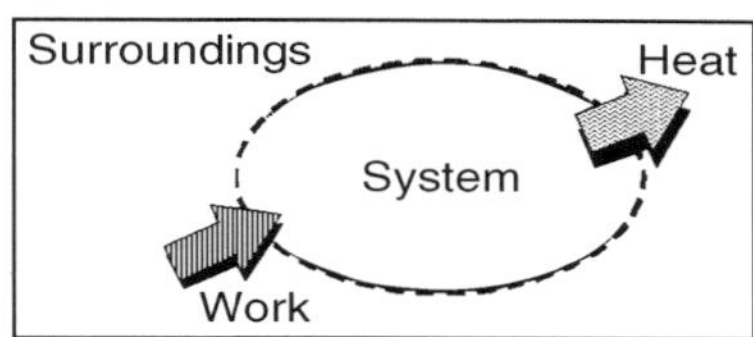

It exchanges energy (heat and work) but not matter with the environment. A greenhouse is an example of a closed system exchanging heat but not work with its environment. Another example is the heat engine shown in Figure below.

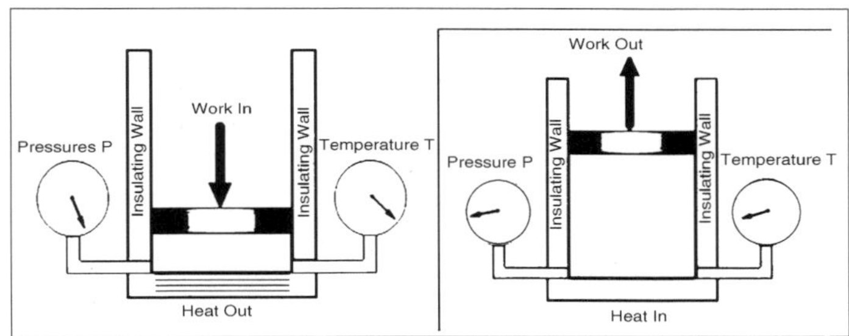

Figure

It is defined as a device that converts heat energy into mechanical energy or more exactly a system which operates with only heat and work passing across its boundaries. As work is done on the gas inside the chamber, the temperature and pressure increase and some heat will be transferred out of the system. When heat is transferred to the system, the gas expands, it does work on the surroundings and the temperature and pressure decrease.

Example: *Piston and cylinder - a closed system*

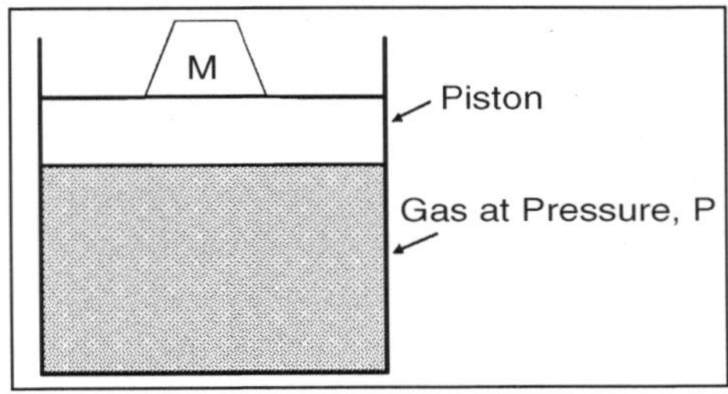

Figure a

OPEN SYSTEM

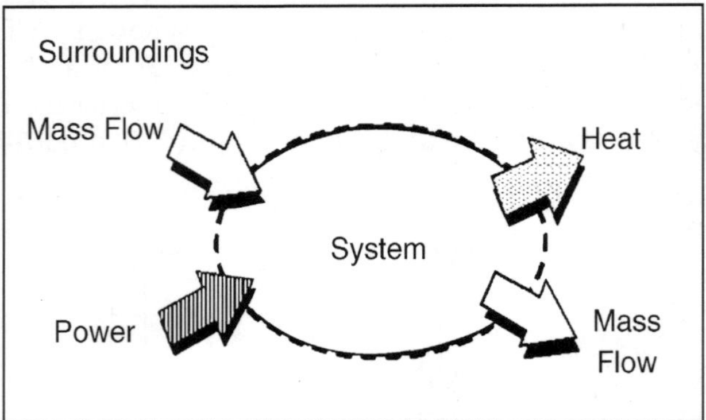

It exchanges energy and matter with the environment. A boundary allowing matter exchange is called permeable. It's possible for an open system to import order and export disorder, locally increasing order.

The Second Law says is that in such a transaction more disorder than order will be created. It does not, however, forbid the creation of pockets of order. What happens is that disorder in the entire system will increase even though individual open systems within it might become more ordered. As shown in Figure below.

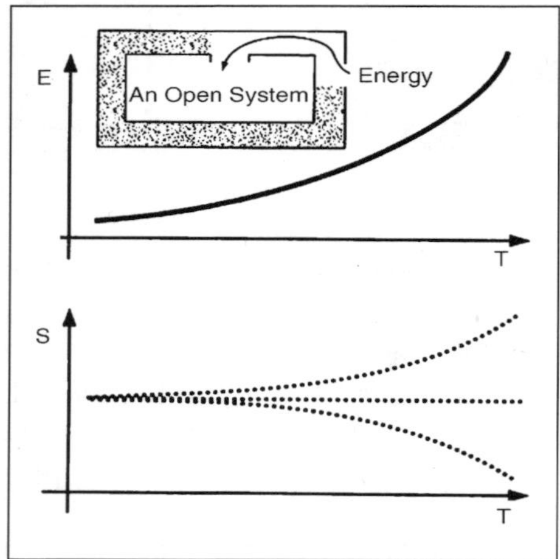

Figure

In a thermodynamically open system, energy can enter the system from the outside environment, thereby increasing the system's total energy, E, over the course of time, t. Such energy flow can lead to an increase, a decrease, or no net change at all in the entropy, S, of the system. Even so, the net entropy of system and its environment would still increase according to the second of thermodynamics.

Example: Photosynthesis in plants as shown in Figure.

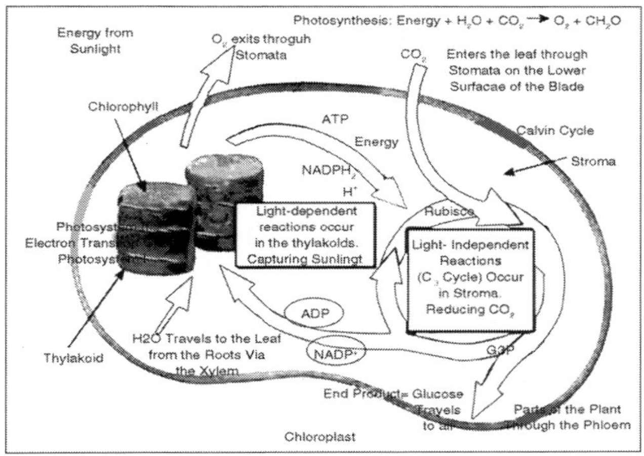

Figure

Infusion of energy and exchange of matter are taking place inside the chloroplast resulting in the production of glucose, which is in a higher energy level. The system becomes nonequilibrium and will decay to the more stable form in the long run.

Example: *The gas turbine engine*

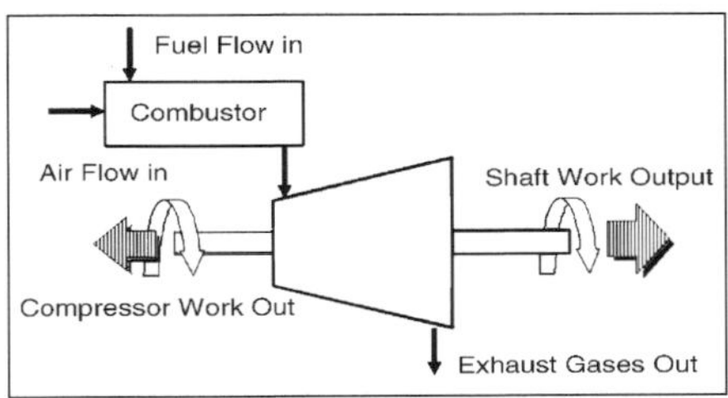

Figure

CONTROL VOLUME

A *control volume* is a fixed region in space chosen for the thermodynamic study of mass and energy balances for flowing systems. The boundary of the control volume may be a real or imaginary envelope. The *control surface* is the boundary of the control volume.

8

Mass of Carbon in Fuel Gases

Modern societies rely on a variety of energy sources to heat homes, propel transportation vehicles, and produce goods for shelter, food, health care, and entertainment. Some of these sources are renewable, whereas others are nonrenewable. A renewable energy source, for example, solar energy, is one that is virtually inexhaustible on the human time scale. A nonrenewable energy source, for example, natural gas, is one that can be either completely consumed (during a lifetime, or during several lifetimes) or depleted to such an extent that it is no longer economical for humankind to obtain it.

About 80 per cent of commercial energy is obtained from three kinds of fuel: oil, coal, and natural gas. These fuels burn in air with the release of energy. They are called fossil fuels because they are believed to have formed from the remains of plants and animals subject to heat and pressure for millions of years.

Natural gas is a mixture of methane (CH_4), 60 to 90 per cent, and smaller amounts of other gaseous hydrocarbons, including ethane (C_2H_6), propane (C_3H_8), and butane (C_4H_{10}). It is valued because it burns hotter and produces less air pollution than other fossil fuels. Completecombustion of a hydrocarbon substance produces carbon dioxide and water.

$$CH_4(g) + 2O_2(g) \rightarrow CO_2(g) + 2H_2O(g) \; \Delta H = 50.1 \text{ kJ/g}$$

In 2001, 2.39 trillion cubic meters of natural gas were consumed worldwide, with estimated remaining reserves of 150 trillion cubic meters. Oil (also referred to as petroleum) is a complex liquid mixture of organic substances, principally of hydrocarbons containing five to sixteen carbon atoms. Most crude oil, once removed from a well, is sent by pipeline to a refinery, where it is distilled to separate it into gasoline, heating oil, diesel oil, and asphalt. The use of catalysts during the refining process increases the yield of gasoline. In 2001, 25.7 billion barrels of oil were used worldwide, with estimated reserves of 1.05 trillion barrels. (One barrel contains 159 liters.)

Coal is the most plentiful fossil fuel, comprising 80 per cent of the fuel reserves of the United States and 90 per cent of those of the world. It is a complex mixture of organic compounds and is anywhere from 30 to 95 per cent carbon by mass. It also contains sulfur compounds. When coal is burned, the sulfur is converted to sulfur dioxide, a troublesome air pollutant.

The description of coal as being of high quality is based on its having a low sulfur content and a high carbon content. Lignite coal (brown coal) has a low carbon content and produces the least energy upon combustion (about 15 kJ/g). Bituminous coal (soft coal) has a higher carbon content and produces more energy. It is the most extensively used coal. Anthracite coal (hard coal) has the highest carbon and heat content (about 30 kJ/g), but supplies of it are limited in most places. In 2001, 4.41 billion metric tons of coal were consumed worldwide, with estimated reserves of 985 billion metric tons. (A metric ton is 1,000 kilograms [2,679 pounds].)

The combustion of fossil fuels produces carbon dioxide gas, a heat-trapping gas. For the past 250 years (since the beginning of the Industrial Revolution), the increased use of fossil fuels has caused the atmospheric concentration of carbon dioxide to increase by a factor of about 25 per cent.

It is now generally believed that this increase has produced higher global temperatures—a phenomenon called the greenhouse effect. Commercial nuclear power is generated by nuclear fission reactions. When slow-moving neutrons strike nuclei of uranium-235 or plutonium-239, these nuclei are split, releasing energy. The energy is used to heat water and drive a turbine, in turn producing electrical energy. Currently nuclear power supplies more than 16 per cent of the world's total electricity. A typical nuclear reactor utilizes uranium oxide, whose uranium content is approximately 3 per cent uranium-235, and 97 per cent uranium-238, by mass. During the fission reaction, the uranium-235 is consumed and fission products form. As the amount of uranium-235 decreases and the amounts of fission products increase, the fission process becomes less efficient.

At some point, the spent nuclear fuel is removed and stored. Some of the radioactive fission products, because of their radioactivity and long half-lives, must be stored securely for thousands of years. Thus, nuclear waste management poses a tremendous challenge.

Scientists hope to someday use controlled nuclear fusion to produce energy. Nuclear fusion, which involves the coming together of light nuclei to form heavier ones, is the process by which stars generate energy. In order for nuclear fusion to occur, the nuclei must have extremely high temperatures. Research has focused on the fusion of deuterium (hydrogen–2) nuclei and tritium (hydrogen–3) nuclei, a process that requires about 50 million degrees Celsius.

The principal renewable energy sources are biomass from crops such as trees and corn, hydropower from flowing rivers, geothermal power from heat stored in Earth, wind energy from the movement of winds, and solar energy from the Sun. Wood is part of an array of plant matter referred to as biomass that can be burned to produce energy. The combustible substances in biomass are primarily carbohydrates (and of these, primarily cellulose). Cellulose, whose simplest or empirical formula is CH_2O, undergoes combustion to form carbon dioxide and water.

$$CH_2O(s) + O_2(g) \rightarrow CO_2(g) + H_2O(g) \quad \Delta H = -14 kj/g$$

Wood fuels continue to be used in the rural areas of developing countries.

Hydroelectric power is a well-developed energy source. Today, hydropower provides about 19 per cent of the world's electricity supply. Because it is a clean, renewable source of energy, hydropower should continue to serve as a vital energy source.

There has been a rapid growth in the use of wind turbines to generate electricity. In 2001 the amount of electricity generated in this way worldwide corresponded to the amount that would have been obtained from burning 15 million barrels of oil. Although this represents only about 0.05 per cent of worldwide energy production in 2001, this fraction will increase.

Solar energy is the most significant and promising renewable energy source. Solar energy is converted to electricity by solar cells (also known as photovoltaic cells). A great deal of solar energy is used currently in what is known as passive heating (which can be directly experienced as the heat gain in a greenhouse caused by sunlight).

TECHNOLOGIES FOR MEASURING COMBUSTION FLUE GASES

OXYGEN

The most ubiquitous technology for measuring combustion flue gases has been the zirconium oxide fuel cell oxygen analyser. This analyser technology was first used in the power generation industry in the early 1970s, but the technology has transferred to use for any combustion process. All automobiles now use one or more of these sensors for controlling fuel-air ratios, and small engines for lawn mowers, chain saws, etc. will soon be using them. Much has been written about the details of how the Nernstian phenomenon operates[1], and this paper will not review this information.

Fig. ZrO_2 Sensing Cell Mounted to the End of a Probe (0.5–6m long).

The ZrO_2 sensing technology is ideally suited for measuring combustion flue gases for the following reasons:

- The sensing cell generates its own millivolt signal, similar to the way a thermocouple works.
- This raw millivolt signal is inverse and logarithmic, i.e. increasing

greatly with the low O_2 readings typically found in combustion processes. Accuracy actually improves as O_2 levels decrease.

- The sensor is typically heated to 700–750°C, so operation in hot combustion flue gases does not present a problem
- The sensor is robust, and can withstand the sulfur components found in many fuels.
- No sampling system is required. The sensor can be placed directly into the flue gas stream on the end of probe that can be from 0.5m to 6m long. Since the flue gases enter the sensor via passive diffusion, even applications with heavy particulate content are possible with a low rate of filter pluggage.
- Sensors can be calibrated on-line and in-place. Automated calibration is also available.

The in situ ZrO_2 probe results in a point measurement within the flue gas duct, however, and several probes of different lengths may be required in order to get a representative average across large flue gas ducts.

Carbon Monoxide

CO is usually the first combustible gas component to appear when combustion fuel/air ratios start becoming too rich. Desired CO levels in combustion flue gases are typically less than 200 PPM, and infra-red spectroscopy is well suited to measuring at these low levels.

Repeatabilities of better than +/–5PPM are possible, with low interference from H_2O and CO_2. Instrument configurations include:

- Eextractive systems where the flue gases are removed from the duct and cleaned before being placed into a rack-mounted analyser.
- Across duct line-of-sight configurations whereby an infra-red source is mounted on one side of the duct, and a receiver or detector is mounted to the opposite side.

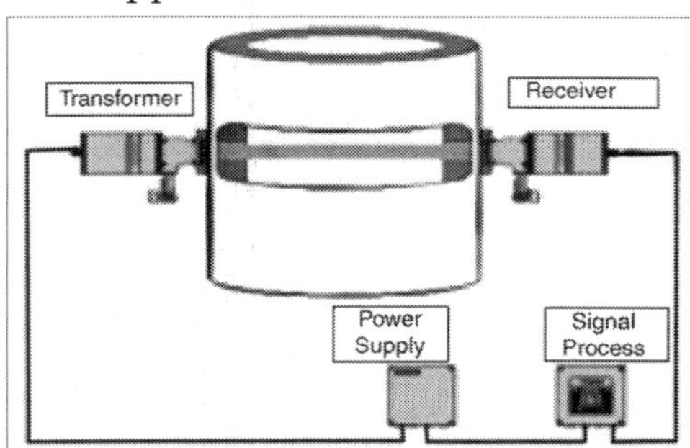

Fig. Typical Across-duct Infra-red Measuring Arrangement.

This line-of –sight method results in an inherent average across the entire duct, so multiple instruments are less likely required to cover a large duct. Conversely, one does not get the granularity of information that an array of point measurement O_2 probes provides.

- *Dual-pass probe*: A modification to the across stack line-of sight method, this arrangement is a dual pass probe where the infra red energy is sent out to a mirror at the end of a hollow pipe, and then reflected back to the source end for analysis. The flue gases are permitted to fill the probe tube through holes or filters.

Any optical technology presents application challenges that need to be considered:

- An extractive system involves transporting and filtering the sample flue gases, removing the moisture, and returning the sample to the process or to a safe vent. This adds considerable cost to the system, and will require significant maintenance attention if there is particulate in the flue gases.
- An across duct line-of –sight system cannot be placed where temperatures are much above 600°C, nor endure high levels of particulate. Thermal growth of the ductwork and vibration can negatively impact the alignment of the source and receiver sides. Also, this type of arrangement cannot undergo a true calibration, since this would involve filling the entire duct with calibration gases.
- A dual-pass probe system has to contend with soiling of the reflecting mirror at the end of the probe. It is possible to conduct an on-line calibration by filling the optical path inside the pipe with calibration gases.

Tunable diode lasers (TDL) have recently come onto the scene, again using spectroscopy, but with a laser source and a diode sensing array. These systems typically use the line-of-sight arrangement across the duct, or a dual pass probe method. This technology is also capable of measuring O_2 in the overtone range, and NO_x. Again, much has been written about the underlying technology, so we will not cover this in this paper. As with the traditional IR systems, a TDL in an across stack line-of-sight arrangement will inherently average across the entire furnace volume, requiring fewer instruments to cover a large duct, but also providing less granularity of the flue gases within the duct. Analysers cannot be challenged with known calibration gases.

APPLICATIONS

Each of the 20 or more burners in a large boiler can be considered as separate processes, each producing its own flue gases. The flue gases in furnaces utilizing burners in a single or opposed-wall firing configuration tend to form up into "columns" that often tend to stay stratified throughout the furnace. Combustion analysers are typically mounted in the "back pass" of the furnace, after the economizer, and it is common for operators to see this stratification when using multiple O_2 probes in these large ducts. It's not uncommon to see differences of one per cent or more across a large furnace.

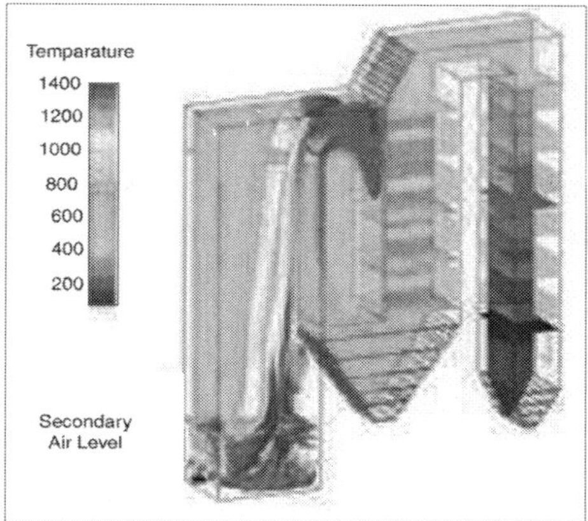

Fig. CFD Depiction of the Flue Gases Passing Through a Wall-fired Furnace

To accommodate this, an arithmetic average of multiple probes is often calculated in the DCS, and used as input to the O_2 control loop.

Forward-thinking operators will use these varying O_2*indications as a diagnostic tool to look for problems in the furnace, such as:*

- Fouled burners
- Sticking sleeve dampers
- ID fan imbalances
- Roping in coal pipes
- Classifier pluggage/coal finess problems
- Coal mill imbalances

Fig. An Array of Oxygen Probes Mounted Vertically Downstream of an Economizer.

The flue gases passing through a tangential-fired furnace do not experience as much stratification of the flue gases, and burner to burner variations are harder to differentiate. Some operators claim to be able to detect corner to corner variations with the O_2 probes in tangential furnaces, however.

Abrasion-Resistance

Furnaces firing solid fuels (particularly coal) can have high levels of fly ash carried with the combustion flue gases. Separate schedule 40 pipes are often used as "abrasive shields" to protect in situ oxygen probes from fly ash erosion. Some operations have discovered, however, that fly ash is often much less abrasive in the hotter zones of the furnace (500–700°C), above the economizer. As the combustion flue gases are cooled through the economizer and air heater, the ash often agglomerates into larger particles that are far more abrasive. A location higher in the furnace can not only minimize abrasion, but also detect stratification better. Rosemount Analytical has developed a heavy-wall probe body that is more cost-effective than traditional abrasion shields, yet endures fly ash erosion well in high areas of coal-fired boilers.

Ideal Probe Placement

Oxygen probes are provided in a wide range of lengths, from 0.5m to 6m, but plant engineers often wonder if a given placement is the optimum. A variable insertion capability has been developed that permits the Instrument Engineer to find the best possible mounting location for a given probe.

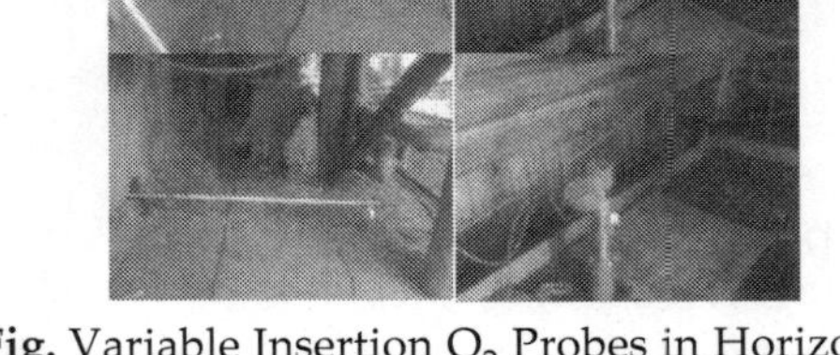

Fig. Variable Insertion O_2 Probes in Horizontal and Vertical Orientations

Air Heater and Other Duct Seal Leaks

Some air heater styles rotate like a revolving door in order to exchange remaining heat from the flue gases to the fresh air being fed to the burners. As the seals in these large rotating structures wear, air will typically migrate over to the flue gas side, elevating the O_2 levels on the flue gas side of the air heater. Any air leak into a furnace negatively affects heat rate efficiency, and also reduces fan capacity used for combustion, limiting boiler capacity.

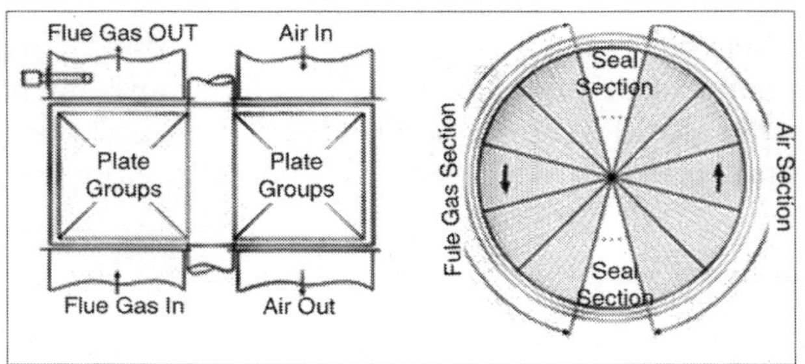

Fig. Rotating Air Heater

New Sensor Developments

Continued research into the ZrO_2 fuel cell technology is yielding new capabilities. It was previously mentioned that the millivolt output of these sensing cells is inverse and logarithmic, so lower levels of oxygen results in higher levels of signal. A sensing cell has been developed that will continue outputting increasing voltage as flue gas O_2 levels pass through zero and into reducing conditions.

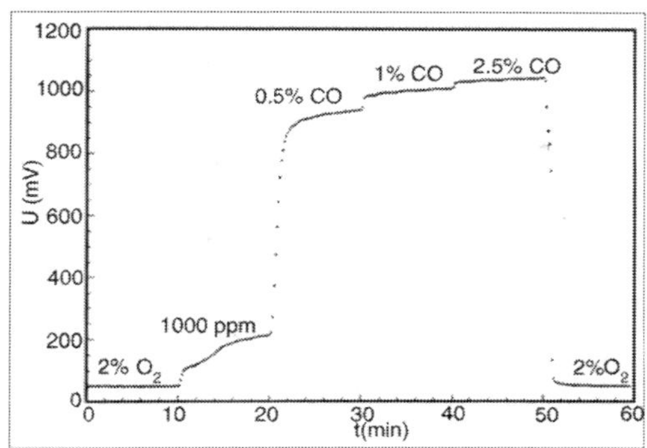

Fig. Millivolt Signal Output from a New ZrO_2 Sensing Cell in Oxidizing and Reducing Conditions.

This has become a tool for processes that periodically pass into reducing conditions, providing an indication of the level of oxygen deficiency during these events, and informing the operator if recovery measures being taken are being effective. Continued research into the ZrO_2 fuel cell technology is yielding new capabilities including a ZrO_2cell that measures CO. Eight test sites have been established at N. American power plants with promising results.

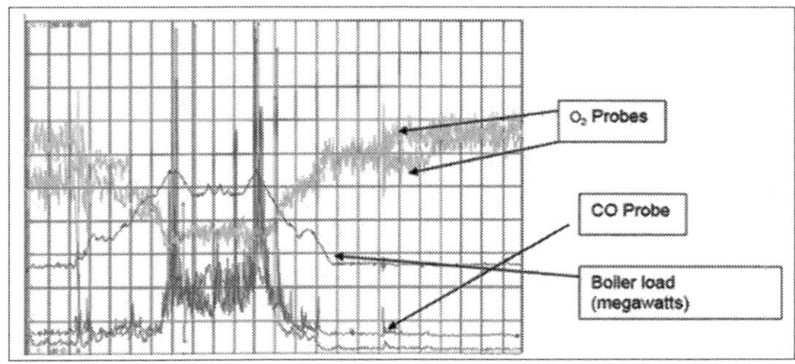

Fig. DCS Trace Depicting PPM Reading from Alpha CO Probe.

THE ANALYSIS OF COMBUSTION FUEL GAS

Combustion flue gas analysis has been used by power plant operators for decades as a method of optimizing the fuel/air ratio. By measuring the amount of excess oxygen and/or CO in the flue gases resulting from combustion, plant operators can ensure that their facility works at the best heat rate efficiency and avoids unnecessary NO_x and greenhouse gas emissions.

The theoretical ideal, or the stoichiometric point, is that in which all fuel is reacted with available oxygen in the combustion air, with none of any of the two reactants left over.

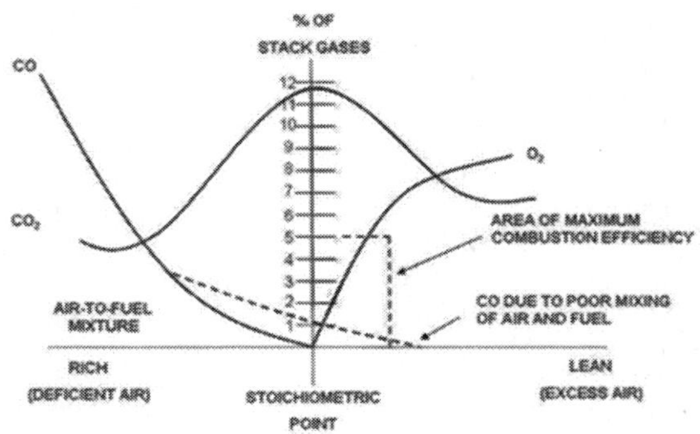

Fig. Key Flue Gas Measurements Relating to Ideal Combustion Stoichiometry

Operating furnaces never attain this ideal, however, and the best operating point usually will result in 1–3 per cent excess air, and 0–200PPM of CO. This optimum operating point is different for every furnace, and also varies for differing loads, or firing rates.

A higher firing rate induces greater turbulence through the burner(s), providing better mixing of fuel and air, and enabling operation with a lower excess O_2 before unburned fuel (represented by CO) appears, or "breaks through".

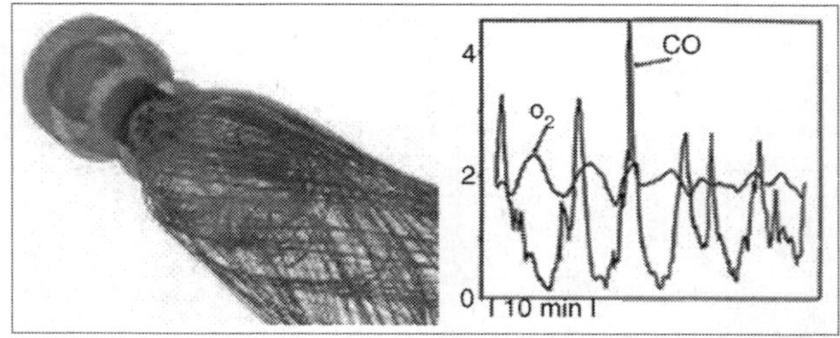

Fig. (Left): CFD Depiction of the Turbulent Mixing of Fuel and Air Through a Burner. (Right): DCS Trend Depicting the Relationship of O_2 and CO Indications at CO Breakthrough Point.

Again, this ideal O_2 operating point will vary with firing rate, so a function generator is usually developed from test data to assign the ideal O_2 control point based upon an index of firing rate, such as fuel flow or steam flow.

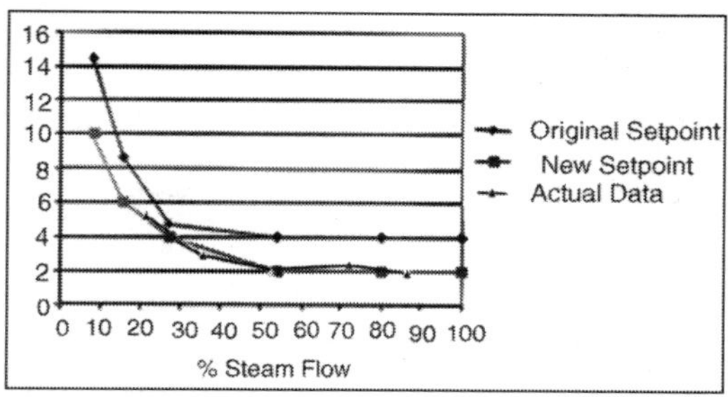

Fig. A Typical Function Generator Depicting the Optimum Flue Gas O_2 Level at Different Steam Flows (firing rates).

This curve should be reestablished from time to time as burners wear, and other furnace conditions change over time. The curve for burners using natural gas and light oil fuels will tend to remain valid for long periods of time (years). Burners firing solid fuels such as coal, petroleum coke, or pellitized biofuels will experience more frequent pluggage and other degradation in the burners and fuel delivery systems, and will benefit from more frequent reestablishment of this curve. Large furnace operators will typically dynamically control oxygen to the optimal level via the distributed control system. Control of CO is more difficult, since target levels are usually in the PPM range, and making fan or damper adjustments small enough to control at these low levels is difficult. Many operators will make manual adjustments based upon the CO signal, or use the measurement as a feed forward signal to adjust the O_2 control setpoint upwards or downwards.

NEW GOALS

The traditional goal of achieving best combustion efficiency is sometimes being modified to accommodate two other goals:

1. Minimizing the thermal NO_x produced through the burner. O_2 levels and flame temperatures are key indicators to the production of NO_x.

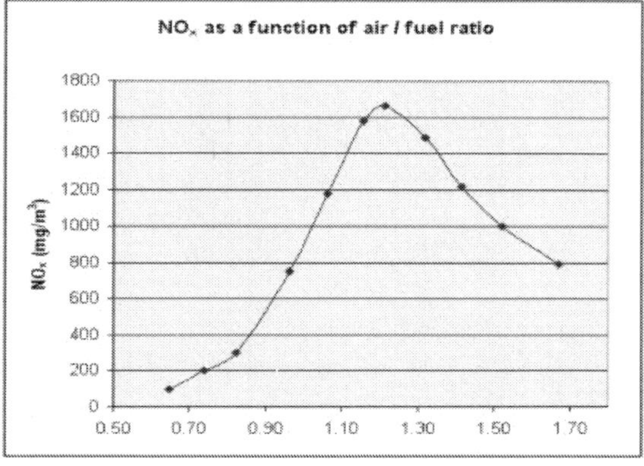

Fig. NOx as a Function of Flue Gas Excess O_2 Relationship of NOx Production

One operating strategy to produce less NO_x uses staged combustion, whereby a cooler fuel-rich combustion is established at the burner. Overfire air is then added higher in the furnace to complete the combustion. This results in less heat and oxygen passing through the burner, and less NO_x produced. Advanced control strategies utilizing neural nets are often implemented to find the optimum air settings to minimize thermal NO_x production. Another NO_x reduction strategy is flue gas recirculation, where a certain amount of flue gas is mixed with the normal air used for combustion. An O_2 probe mounted after this mixing valve can be used to control final O_2 going to the burner, resulting in a cooler flame that produces less NO_x.

2. *Slag prevention*: Flux sensors provide good information about soot and slag buildup, but close attention to combustion analysers can provide another indication of slag formation. Fly ash fusion temperatures are usually affected by the amount of excess O_2 in the flue gases, and some operators run with an O_2 setpoint that has been established to prevent slag.

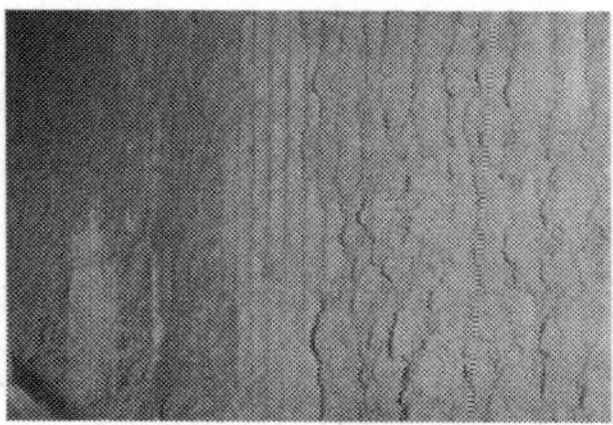

Fig. Slag Formation on Boiler Tubes

9

Measurement of Heat Capacity

The heat capacity of most systems is not a constant. Rather, it depends on the state variables of the thermodynamic system under study. In particular it is dependent on temperature itself, as well as on the pressure and the volume of the system, and the ways in which pressures and volumes have been allowed to change while the system has passed from one temperature to another. The reason for this is that pressure-volume work done to the system raises its temperature by a mechanism other than heating, while pressure-volume work done *by* the system absorbs heat without raising the system's temperature.

Different measurements of heat capacity can therefore be performed, most commonly at constant pressure and constant volume. The values thus measured are usually subscripted (by p and V, respectively) to indicate the definition. Gases and liquids are typically also measured at constant volume. Measurements under constant pressure produce larger values than those at constant volume because the constant pressure values also include heat energy that is used to do work to expand the substance against the constant pressure as its temperature increases. This difference is particularly notable in gases where values under constant pressure are typically 30 to 66.7 per cent greater than those at constant volume.

The specific heat capacities of substances comprising molecules are not fixed constants and vary somewhat depending on temperature. Accordingly, the temperature at which the measurement is made is usually also specified. Examples of two common ways to cite the specific heat of a substance are as follows:

- Water (liquid): c_p = 4.1855 [J/(g·K)] (15 °C, 101.325 kPa)
- Water (liquid): C_vH = 74.539 J/(mol·K) (25 °C)

For liquids and gases, it is important to know the pressure to which given heat-capacity data refer. Most published data are given for standard pressure. However, quite different standard conditions for temperature and pressure have been defined by different organizations. The International Union of Pure and Applied Chemistry changed its recommendation from one atmosphere to the round value 100 kPa (≈750.062 Torr).

CONCEPTS OF HEAT CAPACITY

When heat is transferred to an object, the temperature of the object increases. When heat is removed from an object, the temperature of the object decreases. The relationship between the heat (q) that is transferred and the change in temperature (ΔT) is

$$q = C\,\Delta T = C\,(T_f - T_i)$$

The proportionality constant in this equation is called the heat capacity (C). The heat capacity is the amount of heat required to raise the temperature of an object or substance one degree. The temperature change is the difference between the final temperature (T_f) and the initial temperature (T_i).

A calorimeter is an experimental device in which a chemical reaction or physical process takes place. The calorimeter is well-insulated so that, ideally, no heat enters or leaves the calorimeter from the surroundings. For this reason, any heat liberated by the reaction or process being studied must be picked up by the calorimeter and other substances in the calorimeter.

A thermometer is typically inserted in the calorimeter to measure the change in temperature that results from the reaction or physical process. A stirrer is employed to keep the contents of the calorimeter well-mixed and to ensure uniform heating.

Some water and is equipped with a thermometer, a stirrer, and a heating element. When activated, an electric current is passed through the heating element to generate heat, which is transferred to the calorimeter. In this simulation, you can set the heating rate (units of J sec^{-1}) and the heating time (units of sec).

ADIABATIC CALORIMETERS

An adiabatic calorimeter is a calorimeter used to examine a runaway reaction. Since the calorimeter runs in an adiabatic environment, any heat generated by the material sample under test causes the sample to increase in temperature, thus fuelling the reaction.

No adiabatic calorimeter is truly adiabatic-some heat will be lost by the sample to the sample holder. A mathematical correction factor, known as the phi-factor, can be used to adjust the calorimetric result to account for these heat losses. The phi-factor is the ratio of the thermal mass of the sample and sample holder to the thermal mass of the sample alone.

Examples of adiabatic calorimeters are:-

- THT EV-Accelerating Rate Calorimeter
- HEL Phi-Tec.

A Simple Dewar Flask

Systag FlexyTSC a successor of their SIKAREX unit-the electronics of which could be used to apply a feedback system to heat the sample holder to give a result closer to true adiabaticy, however as the sample holder is an open ended glass tube, one soon loses the sample as a great deal of smoke.

REACTION CALORIMETERS

A reaction calorimeter is a calorimeter in which a chemical reaction is initiated within a closed insulated container. Reaction heats are measured and the total heat is obtained by integrating heatflow versus time. This is the standard used in industry to measure heats since industrial processes are engineered to run at constant temperatures. Reaction calorimetry can also be used to determine maximum heat release rate for chemical process engineering and for tracking the global kinetics of reactions. There are four main methods for measuring the heat in reaction calorimeter:

Heat Flow of Calorimetry

The cooling/heating jacket controls either the temperature of the process or the temperature of the jacket. Heat is measured by monitoring the temperature difference between heat transfer fluid and the process fluid. In addition fill volumes (*i.e.* wetted area), specific heat, heat transfer coefficient have to be determined to arrive at a correct value. It is possible with this type of calorimeter to do reactions at reflux, although the accuracy is not as good.

Heat Balance of Calorimeter

The cooling/heating jacket controls the temperature of the process. Heat is measured by monitoring the heat gained or lost by the heat transfer fluid.

Power Compensation

Power compensation uses a heater placed within the vessel to maintain a constant temperature. The energy supplied to this heater can be varied as reactions require and the calorimetry signal is purely derived from this electrical power.

Constant Flux of Calorimetry

Constant flux calorimetry (or COFLUX as it is often termed) is derived from heat balance calorimetry and uses specialized control mechanisms to maintain a constant heat flow (or flux) across the vessel wall.

CALCULATION FROM FIRST PRINCIPLES

The path integral Monte Carlo method is a numerical approach for determining the values of heat capacity, based on quantum dynamical principles. However, good approximations can be made for gases in many states using simpler methods outlined below. For many solids composed of relatively heavy atoms (atomic number > iron), at non-cryogenic temperatures, the heat capacity at room temperature approaches $3R = 24.94$ joules per kelvin per mole of atoms. Low temperature approximations for both gases and solids at temperatures less than their characteristic Einstein temperatures or Debye temperatures can be made by the methods of Einstein and Debye.

THERMODYNAMIC RELATIONS AND DEFINITION OF HEAT CAPACITY

The internal energy of a closed system changes either by adding heat to the system or by the system performing work. Written mathematically we have

$$dU = \delta Q - \delta W.$$

For work as a result of an increase of the system volume we may write,

If the heat is added at constant volume, then the second term of this relation vanishes and one readily obtains

This defines the *heat capacity at constant volume,* C_V. Another useful quantity is the *heat capacity at constant pressure,* C_P. With the *enthalpy* of the system given by

$$H = U + PV$$

our equation for dU changes to;

and therefore, at constant pressure, we have;

RELATION BETWEEN HEAT CAPACITIES

Measuring the heat capacity, sometimes referred to as specific heat, at constant volume can be prohibitively difficult for liquids and solids. That is, small temperature changes typically require large pressures to maintain a liquid or solid at constant volume implying the containing vessel must be nearly rigid or at least very strong. Instead it is easier to measure the heat capacity at constant pressure (allowing the material to expand or contract freely) and solve for the heat capacity at constant volume using mathematical relationships derived from the basic thermodynamic laws. Starting from the fundamental Thermodynamic Relation one can show

$$C_p - C_V = T\left(\frac{\partial p}{\partial T}\right)_{V,N}\left(\frac{\partial V}{\partial T}\right)_{p,N}$$

where the partial derivatives are taken at constant volume and constant number of particles, and constant pressure and constant number of particles, respectively.

This can also be rewritten;

$$C_p - C_V = VT\frac{\alpha^2}{\beta_T}$$

where α

is the coefficient of thermal expansion,

is the isothermal compressibility.

The heat capacity ratio or adiabatic index is the ratio of the heat capacity at constant pressure to heat capacity at constant volume.

It is sometimes also known as the isentropic expansion factor.

Ideal Gas

For an ideal gas, evaluating the partial derivatives above according to the equation of state;

$C_V = nR$

$$pV = nRT$$

the relation can be found to reduce to;

where n is number of moles of gas in the thermodynamic system under consideration, and R is the universal gas constant. Dividing through by n, this equation reduces simply to Mayer's relation,

$$C_{p,m} - C_{v,m} = R$$

where $C_{p,m}$ and $C_{v,m}$ are intensive property heat capacities expressed on a per mole basis at constant pressure and constant volume, respectively.

Specific Heat Capacity

The specific heat capacity of a material on a per mass basis is;

$$c = \frac{\partial C}{\partial m},$$

which in the absence of phase transitions is equivalent to;

$$c = E_m = \frac{C}{m} = \frac{C}{\rho V},$$

where

C is the heat capacity of a body made of the material in question,
m is the mass of the body,
V is the volume of the body, and
$\rho = \frac{m}{V}$ is the density of the material.

For gases, and also for other materials under high pressures, there is need to distinguish between different boundary conditions for the processes under consideration (since values differ significantly between different conditions). Typical processes for which a heat capacity may be defined include isobaric (constant pressure, $dp = 0$) or isochoric (constant volume,) processes. The corresponding specific heat capacities are expressed as;

$$c_V = \left(\frac{\partial C}{\partial m}\right)_V .$$

From the results, dividing through by the mass gives the relation;

$$c_p - c_V = \frac{\alpha^2 T}{\rho \beta_T} .$$

A related parameter to c is CV^{-1} , the volumetric heat capacity.

In engineering practice, c_V for solids or liquids often signifies a volumetric heat capacity, rather than a constant-volume one. In such cases, the mass-specific heat capacity (specific heat) is often explicitly written with the subscript m, as c_m .

Of course, from the above relationships, for solids one writes;

$$c_m = \frac{C}{m} = \frac{c_{volumetric}}{\rho} .$$

For pure homogeneous chemical compounds with established molecular or molar mass or a molar quantity is established, heat capacity as an intensive property can be expressed on a per mole basis instead of a per mass basis by the following equations analogous to the per mass equations:

$C_{p,m} = \left(\frac{\partial C}{\partial n}\right)_p$ = molar heat capacity at constant pressure

$$C_{V,m} = \left(\frac{\partial C}{\partial n}\right)_V$$ = molar heat capacity at constant volume

where n = number of moles in the body or thermodynamic system. One may refer to such a *per mole* quantity as molar heat capacity to distinguish it from specific heat capacity on a per mass basis.

Heat Capacity of Polytropic

The polytropic heat capacity is calculated at processes if all the thermodynamic properties (pressure, volume, temperature) change;

$$C_{i,m} = \left(\frac{\partial C}{\partial n}\right)$$ = molar heat capacity at polytropic process.

The most important polytropic processes run between the adiabatic and the isotherm functions, the polytropic index is between 1 and the adiabatic exponent (γ or κ)

HEAT CAPACITY OF DIMENSIONLESS

The dimensionless heat capacity of a material is;

S / Nk

$$C^* = \frac{C}{nR} = \frac{C}{Nk}$$

where

C is the heat capacity of a body made of the material in question (J/K)
n is the amount of substance in the body (mol)
R is the gas constant (J/(K·mol))
N is the number of molecules in the body. (dimensionless)
k is Boltzmann's constant (J/(K·molecule)).

In the ideal gas article, dimensionless heat capacity C^* v is expressed as $\hat{c}$, and is related there directly to half the number of degrees of freedom per particle. This holds true for quadratic degrees of freedom, a consequence of the equipartition theorem. More generally, the dimensionless heat capacity relates the logarithmic increase in temperature to the increase in the dimensionless entropy per particle , measured in nats.

$$C^* = \frac{dS^*}{d\ln T}$$

Alternatively, using base 2 logarithms, C^* relates the base-2 logarithmic increase in temperature to the increase in the dimensionless entropy measured in bits.

HYDROGEN AS A FUEL FOR HEAT OR ELECTRICITY

Water is perhaps the most important substance to life on Earth. It is a simple compound made from the two elements hydrogen (H) and oxygen (O), and each molecule of water consists of two hydrogen atoms and one oxygen atom. Thus we write the chemical formula for water as "H_2O".

Hydrogen itself is also a very important element in the universe. For example, it is the fuel for the Sun, which generates power by fusing (combining) hydrogen atoms into a helium in a process call nuclear fusion. Because it can be obtained from water, as this project demonstrates, the German's call hydrogen "wasserstoff", which literally means "water stuff". Suppose that you just happen to have some pure hydrogen gas on hand, stored in a container. The hydrogen gas consists of H_2 molecules zipping around in a container (hydrogen atoms like to bond together into H_2 molecules). If there also happens to be oxygen gas around (O_2), and there is always plenty oxygen in the air (air consists of about 20 per cent oxygen), then the oxygen can react violently with the hydrogen gas, such that the hydrogen *burns,* or *combusts,* with the oxygen to form water and heat, according to the chemical reaction

$$2H_2 + O_2 \rightarrow 2\ H_2O + \text{energy (heat)}.$$

Therefore, if you have some hydrogen, you can burn it for fuel to generate heat!

Generating heat, however, is not always the best thing to do, because *entropy,* which may be thought of as molecular *disorder,* is created when heat is generated, and that can limit the efficiency of devices that use that heat energy to do useful work (For more info on entropy, see the section on entropy in our Energy Physics Primer). Fortunately, there exists a device called a *fuel cell,* which can chemically combine hydrogen with oxygen to make electricity. After you complete this project, you may want to also want to cover our project on exploring fuel cells.

Fuel cells can also accomplish what this project demonstrates - electrolysis - which generates hydrogen from water.

Carbon-based Fuels

Notice from the equation above that, unlike burning a fossil (carbon-based) fuel such as coal, burning hydrogen doesn't produce any byproducts except water, which is environmentally benign. Hydrogen burns clean!

Burning fossil fuels on the other hand, always results in carbon monoxide (CO) and/or carbon dioxide (CO_2), which is produced when the carbon atoms combine with oxygen. These compounds are now considered pollutants because they are greenhouse gases - that is, they help trap heat near the Earth's surface, causing the Earth's surface temperature to rise, i.e. global warming. Burning fossil fuels such as coal also usually releases other pollutants, including sulfur dioxide, mercury, and uranium to the atmosphere, because these substances are usually present to varying extent in fossil fuels. But if we can obtain hydrogen without producing greenhouse gases or these other pollutants, then hydrogen is a better fuel to use than fossil fuels. Many people are now hopeful that a "hydrogen-economy" will soon replace our fossil fuel economy.

Obstacles to a Hydrogen-economy

There are two obstacles to a hydrogen-economy.

- It takes alot of volume (or energy) to store hydrogen - usually five times or so the volume, at reasonable pressures, needed to store an equivalent amount of energy with gasoline. One company that has made headway on solving this problem, however, is Dynetek (www.dyneteck.com).
- There is no hydrogen infrastructure: Making the transition to a hydrogen economy might mean having to scrap the fossil fuel infrastructure that we have already developed. One company that has made progress on refueling equipment is Stuart Energy (www.stuartenergy.com).

Both of these problems might be surmounted by using synthetic fuels. For example, it is possible, using a catalyst, to combine water, carbon dioxide (extracted from the air), and renewable electricity to make fuels such as methanol, a carbon-based fuel. When this fuel is burned, water and carbon dioxide are produced. But because the carbon dioxide used initially to make the fuel was extracted from the air, the cycle is closed with respect to both water and carbon dioxide, and so won't contribute excess carbon dioxide to the atmosphere. Fuel cells can already use such fuels (either by extracting the hydrogen from the fuel prior to the fuel cell, or even directly in certain types of fuel cells).

10

Specific Heats Energy

The internal energies and entropies of substances (mostly ideal gases) at some length. Unfortunately, these quantities cannot be directly measured. Instead, they must be inferred from other information. The thermodynamic property of substances which is the easiest to measure is, of course, the heat capacity, or specific heat. In fact, once the variation of the specific heat with temperature is known, both the internal energy and entropy can be easily reconstructed via

$$E(T, V) = v\int_0^T c_V(T,V)dT + E(0,V),$$

$$S(T,V) = v\int_0^T \frac{c_V(T,V)}{T}dT.$$

Here, use has been made of $dS = đQ/T$, and the third law of thermodynamics. Clearly, the optimum way of verifying the results of statistical thermodynamics is to compare the theoretically predicted heat capacities with the experimentally measured values.

Classical physics, in the guise of the equipartition theorem, says that each independent degree of freedom associated with a quadratic term in the energy possesses an average energy $(1/2)kT$ in thermal equilibrium at temperature T. Consider a substance made up of N molecules. Every molecular degree of freedom contributes $(1/2)N\,k\,T$, or $(1/2)\,vRT$, to the mean energy of the substance (with the tacit proviso that each degree of freedom is associated with a quadratic term in the energy). Thus, the contribution to the molar heat capacity at constant volume (we wish to avoid the complications associated with any external work done on the substance) is,

$$\frac{1}{v}\left(\frac{\partial E}{\partial T}\right)_V = \frac{1}{v}\frac{\partial[(1/2)vRT]}{\partial T} = \frac{1}{2}R,$$

per molecular degree of freedom. The total classical heat capacity is therefore,

$$c_V = \frac{g}{2}R,$$

where g is the number of molecular degrees of freedom. Since large complicated molecules clearly have very many more degrees of freedom than small simple molecules, the above formula predicts that the molar heat capacities of substances made up of the former type of molecules should greatly exceed those of substances made up of the latter. In fact, the experimental heat capacities of substances containing complicated molecules are generally greater than those of substances containing simple molecules, but by nowhere near the large factor predicted by previous equation.

This equation also implies that heat capacities are temperature independent. In fact, this is not the case for most substances. Experimental heat capacities generally increase with increasing temperature. These two experimental facts pose severe problems for classical physics. Incidentally, these problems were fully appreciated as far back as 1850. Stories that physicists at the end of the nineteenth century thought that classical physics explained absolutely everything are largely apocryphal.

The equipartition theorem (and the whole classical approximation) is only valid when the typical thermal energy kT greatly exceeds the spacing between quantum energy levels. Suppose that the temperature is sufficiently low that this condition is not satisfied for one particular molecular degree of freedom. In fact, suppose that kT is much less than the spacing between the energy levels.

In this situation the degree of freedom only contributes the ground-state energy, E_0, say, to the mean energy of the molecule. The ground-state energy can be a quite complicated function of the internal properties of the molecule, but is certainly not a function of the temperature, since this is a collective property of all molecules. It follows that the contribution to the molar heat capacity is,

$$\frac{1}{v}\left(\frac{\partial[N\,E_0]}{\partial T}\right)_v = 0.$$

Thus, if kT is much less than the spacing between the energy levels then the degree of freedom contributes nothing at all to the molar heat capacity. We say that this particular degree of freedom is frozen out. Clearly, at very low temperatures just about all degrees of freedom are frozen out. As the temperature is gradually increased, degrees of freedom successively "kick in," and eventually contribute their full (1/2) R to the molar heat capacity, as kT approaches, and then greatly exceeds, the spacing between their quantum energy levels. We can use these simple ideas to explain the behaviours of most experimental heat capacities.

To make further progress, we need to estimate the typical spacing between the quantum energy levels associated with various degrees of freedom. We can do this by observing the frequency of the electromagnetic radiation emitted and absorbed during transitions between these energy levels. If the typical spacing between energy levels is DE then transitions between the various levels are associated with photons of frequency v, where hv = DE. We can define an effective temperature of the radiation via $hv = kT_{rad}$. If $T >> T_{rad}$ then kT >> DE, and the degree of freedom makes its full contribution to the heat capacity. On the other hand, if $T << T_{rad}$ then kT << DE, and the degree of freedom is frozen out. The "temperatures" of various different types of radiation. It is clear that degrees of freedom which give rise to emission or absorption of radio or microwave radiation contribute their full (1/2)R to the molar heat capacity at room temperature.

Degrees of freedom which give rise to emission or absorption in the visible, ultraviolet, X–ray, or g–ray regions of the electromagnetic spectrum are frozen out at room temperature. Degrees of freedom which emit or absorb infrared radiation are on the border line.

Table. Effective "Temperatures" of Various Types of Electromagnetic Radiation

Radiation type	Frequency (Hz)	T_{rad} (°K)
Radio	$< 10^9$	< 0.05
Microwave	$10^9 - 10^{11}$	< 0.05
Infrared	$10^{11} - 10^{14}$	5. 5000
Visible	5×10^{14}	2×10^4
Ultraviolet	$10^{15} - 10^{17}$	5×10^4 5×10^5
X-ray	$10^{17} - 10^{20}$	5×10^6 5×10^9
g-ray	$> 10^{20}$	$> 5 \times 10^9 3$

HEAT ENGINES

Thermodynamics was invented, almost by accident, in 1825 by a young French engineer called Sadi Carnot who was investigating the theoretical limitations on the efficiency of steam engines. Although we are not particularly interested in steam engines, nowadays, it is still highly instructive to review some of Carnot's arguments.

We know, by observation, that it is possible to do mechanical work w upon a device M, and then to extract an equivalent amount of heat q, which goes to increase the internal energy of some heat reservoir. (Here, we use small letters w and q to denote intrinsically positive amounts of work and heat, respectively.)

An example of this is Joule's classic experiment by which he verified the first law of thermodynamics: a paddle wheel is spun in a liquid by a falling weight, and the work done by the weight on the wheel is converted into heat, and absorbed by the liquid.

Carnot's question was this: is it possible to reverse this process and build a device, called a heat engine, which extracts heat energy from a reservoir and converts it into useful macroscopic work? For instance, is it possible to extract heat from the ocean and use it to run an electric generator?

There are a few caveats to Carnot's question. First of all, the work should not be done at the expense of the heat engine itself, otherwise the conversion of heat into work could not continue indefinitely. We can ensure that this is the case if the heat engine performs some sort of cycle, by which it periodically returns to the same macrostate, but, in the meantime, has extracted heat from the reservoir and done an equivalent amount of useful work. Furthermore, a cyclic process seems reasonable because we know that both steam engines and internal combustion engines perform continuous cycles.

The second caveat is that the work done by the heat engine should be such as to change a single parameter of some external device (e.g., by lifting a weight) without doing it at the expense of affecting the other degrees of freedom, or the entropy, of that device. For instance, if we are extracting heat from the ocean to generate electricity, we want to spin the shaft of the electrical generator without increasing the generator's entropy; i.e., causing the generator to heat up or fall to bits.

Let us examine the feasibility of a heat engine using the laws of thermodynamics. Suppose that a heat engine M performs a single cycle. Since M has returned to its initial macrostate, its internal energy is unchanged, and the first law of thermodynamics tell us that the work done by the engine w must equal the heat extracted from the reservoir q, so

$$w = q.$$

The above condition is certainly a necessary condition for a feasible heat engine, but is it also a sufficient condition? In other words, does every device which satisfies this condition actually work? Let us think a little more carefully about what we are actually expecting a heat engine to do. We want to construct a device which will extract energy from a heat reservoir, where it is randomly distributed over very many degrees of freedom, and convert it into energy distributed over a single degree of freedom associated with some parameter of an external device.

Once we have expressed the problem in these terms, it is fairly obvious that what we are really asking for is a spontaneous transition from a probable to an improbable state, which we know is forbidden by the second law of thermodynamics.

So, unfortunately, we cannot run an electric generator off heat extracted from the ocean, because it is like asking all of the molecules in the ocean, which are jiggling about every which way, to all suddenly jig in the same direction, so as to exert a force on some lever, say, which can then be converted into a torque on the generator shaft.

We know from our investigation of statistical thermodynamics that such a process is possible, in principle, but is fantastically improbable. The improbability of the scenario just outlined is summed up in the second law of thermodynamics. This says that the total entropy of an isolated system can never spontaneously decrease, so

$$\Delta S \geq 0.$$

For the case of a heat engine, the isolated system consists of the engine, the reservoir from which it extracts heat, and the outside device upon which it does work. The engine itself returns periodically to the same state, so its entropy is clearly unchanged after each cycle.

We have already specified that there is no change in the entropy of the external device upon which the work is done. On the other hand, the entropy change per cycle of the heat reservoir, which is at absolute temperature T_1, say, is given by

$$\Delta S_{\text{reservoir}} = \oint \frac{đ}{T_1} = -\frac{q}{T_1},$$

where $đQ$ is the infinitesimal heat absorbed by the reservoir, and the integral is taken over a whole cycle of the heat engine. The integral can be converted into the expression $-q/T_1$ because the amount of heat extracted by the engine is assumed to be too small to modify the temperature of the reservoir (this is the definition of a heat reservoir), so that T_1 is a constant during the cycle. The second law of thermodynamics clearly reduces to

$$\frac{-q}{T_1} \geq 0$$

or, making use of the first law of thermodynamics,

$$\frac{q}{T_1} = \frac{w}{T_1} \leq 0.$$

Since we wish the work w done by the engine to be positive, the above relation clearly cannot be satisfied, which proves that an engine which converts heat directly into work is thermodynamically impossible.

A perpetual motion device, which continuously executes a cycle without extracting heat from, or doing work on, its surroundings, is just about possible according to previous equation. In fact, such a device corresponds to the equality sign in earlier equation which means that it must be completely reversible. In reality, there is no such thing as a completely reversible engine. All engines, even the most efficient, have frictional losses which make them, at least, slightly irreversible.

Thus, the equality sign in previous equation corresponds to an asymptotic limit which reality can closely approach, but never quite attain. It follows that a perpetual motion device is thermodynamically impossible. Nevertheless, the U.S. patent office receives about 100 patent applications a year regarding perpetual motion devices. The British patent office, being slightly less open-minded that its American counterpart, refuses to entertain such applications on the basis that perpetual motion devices are forbidden by the second law of thermodynamics. According to previous equation there is no thermodynamic objection to a heat engine which runs backwards, and converts work directly into heat. This is not surprising, since we know that this is essentially what frictional forces do. Clearly, we have, here, another example of a natural process which is fundamentally irreversible according to the second law of thermodynamics. In fact, the statement it is impossible to construct a perfect heat engine which converts heat directly into work is called Kelvin's formulation of the second law. We have demonstrated that a perfect heat engine, which converts heat directly into work, is impossible.

But, there must be some way of obtaining useful work from heat energy, otherwise steam engines would not operate. Well, the reason that our previous scheme did not work was that it decreased the entropy of a heat reservoir, at some temperature T_1, by extracting an amount of heat q per cycle, without any compensating increase in the entropy of anything else, so the second law of thermodynamics was violated. How can we remedy this situation?

We still want the heat engine itself to perform periodic cycles (so, by definition, its entropy cannot increase over a cycle), and we also do not want to increase the entropy of the external device upon which the work is done. Our only other option is to increase the entropy of some other body. In Carnot's analysis, this other body is a second heat reservoir at temperature T_2. We can increase the entropy of the second reservoir by dumping some of the heat we extracted from the first reservoir into it. Suppose that the heat per cycle we extract from the first reservoir is q_1, and the heat per cycle we reject into the second reservoir is q_2. Let the work done on the external device be w per cycle. The first law of thermodynamics tells us that

$$q_1 = w + q_2.$$

Note that $q_2 < q_1$ if positive (i.e., useful) work is done on the external device. The total entropy change per cycle is due to the heat extracted from the first reservoir and the heat dumped into the second, and has to be positive (or zero) according to the second law of thermodynamics. So,

$$\Delta S = \frac{-q_1}{T_1} + \frac{q_2}{T_2} \geq 0.$$

We can combine the previous two equations to give

$$\frac{-q_1}{T_1} + \frac{q_2 - w}{T_2} \geq 0,$$

or

$$\frac{w}{T_2} \leq q_1\left(\frac{1}{T_2} - \frac{1}{T_2}\right).$$

It is clear that the engine is only going to perform useful work (i.e., w is only going to be positive) if $T_2 < T_1$. So, the second reservoir has to be colder than the first if the heat dumped into the former is to increase the entropy of the Universe more than the heat extracted from the latter decreases it. It is useful to define the efficiency η of a heat engine.

This is the ratio of the work done per cycle on the external device to the heat energy absorbed per cycle from the first reservoir. The efficiency of a perfect heat engine is unity, but we have already shown that such an engine is impossible. What is the efficiency of a realizable engine? It is clear from the previous equation that

$$\eta \equiv \frac{w}{q_2} \leq 1 \frac{T_2}{T_1} = \frac{T_1 - T_2}{T_1}.$$

Note that the efficiency is always less than unity. A real engine must always reject some energy into the second heat reservoir in order to satisfy the second law of thermodynamics, so less energy is available to do external work, and the efficiency of the engine is reduced. The equality sign in the above expression corresponds to a completely reversible heat engine (i.e., one which is quasi-static). It is clear that real engines, which are always irreversible to some extent, are less efficient than reversible engines. Furthermore, all reversible engines which operate between the two temperatures T_1 and T_2 must have the same efficiency,

$$\eta = \frac{T_1 - T_2}{T_1},$$

irrespective of the way in which they operate.

Let us consider how we might construct one of these reversible heat engines. Suppose that we have some gas in a cylinder equipped with a frictionless piston. The gas is not necessarily a perfect gas. Suppose that we also have two heat reservoirs at temperatures T_1 and T_2 (where $T_1 > T_2$). These reservoirs might take the form of large water baths. Let us start off with the gas in thermal contact with the first reservoir. We now pull the piston out very slowly so that heat energy flows reversibly into the gas from the reservoir.

Let us now thermally isolate the gas and slowly pull out the piston some more. During this adiabatic process the temperature of the gas falls (since there is no longer any heat flowing into it to compensate for the work it does on the piston). Let us continue this process until the temperature of the gas falls to T_2.

We now place the gas in thermal contact with the second reservoir and slowly push the piston in. During this isothermal process heat flows out of the gas into the reservoir. We next thermally isolate the gas a second time and slowly compress it some more. In this process the temperature of the gas increases.

We stop the compression when the temperature reaches T_1. If we carry out each step properly we can return the gas to its initial state and then repeat the cycle ad infinitum. We now have a set of reversible processes by which a quantity of heat is extracted from the first reservoir and a quantity of heat is dumped into the second.

We can best evaluate the work done by the system during each cycle by plotting out the locus of the gas in a p–V diagram. The locus takes the form of a closed curve.

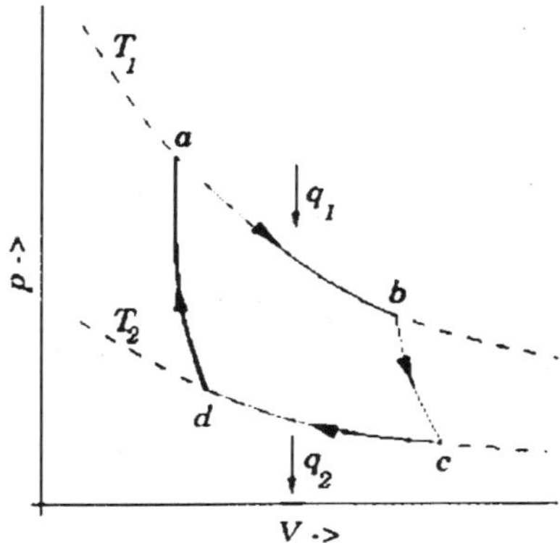

Fig. An ideal Gas Carnot Engine.

The net work done per cycle is the "area" contained inside this curve, since $đW = p\ dV$ [if p is plotted vertically and V horizontally, then $p\ dV$ is clearly an element of area under the curve $p\ (V)$]. The engine we have just described is called a Carnot engine, and is the simplest conceivable device capable of converting heat energy into useful work.

For the specific case of an ideal gas, we can actually calculate the work done per cycle, and, thereby, verify previous equation. Consider the isothermal expansion phase of the gas. For an ideal gas, the internal energy is a function of the temperature alone. The temperature does not change during isothermal expansion, so the internal energy remains constant, and the net heat absorbed by the gas must equal the work it does on the piston. Thus,

$$q_1 = \int_a^b p dV,$$

where the expansion takes the gas from state a to state b. Since $pV = vRT$, for an ideal gas, we have

$$q_1 = \int_a^b v\, RT_1 \frac{dV}{V} = vRT_1 \ln \frac{V_b}{V_a}.$$

Likewise, during the isothermal compression phase, in which the gas goes from state c to state d , the net heat rejected to the second reservoir is

$$q_1 = vRT_2 \ln \frac{V_c}{V_d}.$$

Now, during adiabatic expansion or compression

$$TV^{\gamma-1} = \text{constant}.$$

It follows that during the adiabatic expansion phase, which takes the gas from state b to state c,

$$T_1 V_b^{\gamma-1} = T_2 V_c^{\gamma-1}.$$

Likewise, during the adiabatic compression phase, which takes the gas from state d to state a,

$$T_1 V_a^{\gamma-1} = T_2 V_d^{\gamma-1}.$$

If we take the ratio of the previous two equations we obtain

$$\frac{V_b}{V_a} = \frac{V_c}{V_d}.$$

Hence, the work done by the engine, which we can calculate using the first law of thermodynamics,

$$w = q_1 - q_2,$$

is

$$w = vR\,(T_1 - T_2)\ln\frac{V_b}{V_a}.$$

Thus, the efficiency of the engine is

$$\eta = \frac{w}{q_1} = \frac{T_1 - T_2}{T_1}$$

which, not surprisingly, is exactly the same as previous equation.

The engine described above is very idealized. Of course, real engines are far more complicated than this. Nevertheless, the maximum efficiency of an ideal heat engine places severe constraints on real engines. Conventional power stations have many different "front ends" (e.g., coal fired furnaces, oil fired furnaces, nuclear reactors), but their "back ends" are all very similar, and consist of a steam driven turbine connected to an electric generator.

The "front end" heats water extracted from a local river and turns it into steam, which is then used to drive the turbine, and, hence, to generate electricity.

Finally, the steam is sent through a heat exchanger so that it can heat up the incoming river water, which means that the incoming water does not have to be heated so much by the "front end." At this stage, some heat is rejected to the environment, usually as clouds of steam escaping from the top of cooling towers. We can see that a power station possesses many of the same features as our idealized heat engine.

There is a cycle which operates between two temperatures. The upper temperature is the temperature to which the steam is heated by the "front end," and the lower temperature is the temperature of the environment into which heat is rejected. Suppose that the steam is only heated to 100 °C (or 373° K), and the temperature of the environment is 15 °C (or 288 °K). It follows from Eq. that the maximum possible efficiency of the steam cycle is

$$\eta = \frac{373 - 288}{373} \simeq 0.23.$$

So, at least 77 per cent of the heat energy generated by the "front end" goes straight up the cooling towers! Not be surprisingly, commercial power stations do not operate with 100 °C steam. The only way in which the thermodynamic efficiency of the steam cycle can be raised to an acceptable level is to use very hot steam (clearly, we cannot refrigerate the environment). Using 400 °C steam, which is not uncommon, the maximum efficiency becomes

$$\eta = \frac{673 - 288}{673} \simeq 0.57,$$

which is more reasonable. In fact, the steam cycles of modern power stations are so well designed that they come surprisingly close to their maximum thermodynamic efficiencies.

SPECIFIC HEATS OF SOLIDS

Consider a simple solid containing N atoms. Now, atoms in solids cannot translate (unlike those in gases), but are free to vibrate about their equilibrium positions. Such vibrations are called lattice vibrations, and can be thought of as sound waves propagating through the crystal lattice. Each atom is specified by three independent position coordinates, and three conjugate momentum coordinates.

Let us only consider small amplitude vibrations. In this case, we can expand the potential energy of interaction between the atoms to give an expression which is quadratic in the atomic displacements from their equilibrium positions. It is always possible to perform a normal mode analysis of the oscillations. In effect, we can find $3N$ independent modes of oscillation of the solid. Each mode has its own particular oscillation frequency, and its own particular pattern of atomic displacements. Any general oscillation can be written as a linear combination of these normal modes. Let q_i be the (appropriately normalized) amplitude of the ith normal mode, and $p_i$the momentum conjugate to this coordinate. In normal mode coordinates, the total energy of the lattice vibrations takes the particularly simple form,

$$E = \frac{1}{2}\sum_{i=1}^{3N}(p_i^2 + \omega_i^2 q_i^2),$$

where w$_i$ is the (angular) oscillation frequency of the ith normal mode. It is clear that in normal mode coordinates, the linearized lattice vibrations are equivalent to $3N$ independent harmonic oscillators (of course, each oscillator corresponds to a different normal mode).

The typical value of w_i is the (angular) frequency of a sound wave propagating through the lattice. Sound wave frequencies are far lower than the typical vibration frequencies of gaseous molecules.

In the latter case, the mass involved in the vibration is simply that of the molecule, whereas in the former case the mass involved is that of very many atoms (since lattice vibrations are non-localized).

The strength of interatomic bonds in gaseous molecules is similar to those in solids, so we can use the estimate $w \sim \sqrt{k/m}$ (k is the force constant which measures the strength of interatomic bonds, and m is the mass involved in the oscillation) as proof that the typical frequencies of lattice vibrations are very much less than the vibration frequencies of simple molecules.

It follows from $DE = \hbar\omega$ that the quantum energy levels of lattice vibrations are far more closely spaced than the vibrational energy levels of gaseous molecules. Thus, it is likely (and is, indeed, the case) that lattice vibrations are not frozen out at room temperature, but, instead, make their full classical contribution to the molar specific heat of the solid.

If the lattice vibrations behave classically then, according to the equipartition theorem, each normal mode of oscillation has an associated mean energy kTin equilibrium at temperature T [(1/2)kT resides in the kinetic energy of the oscillation, and (1/2)kT resides in the potential energy]. Thus, the mean internal energy per mole of the solid is,

$$\overline{E} = 3NkT = 3vRT.$$

It follows that the molar heat capacity at constant volume is,

$$c_V = \frac{1}{v}\left(\frac{\partial \overline{E}}{\partial T}\right)_V = 3R$$

for solids. This gives a value of 24.9 joules/mole/degree. In fact, at room temperature most solids (in particular, metals) have heat capacities which lie remarkably close to this value. This fact was discovered experimentally by Dulong and Petite at the beginning of the nineteenth century, and was used to make some of the first crude estimates of the molecular weights of solids (if we know the molar heat capacity of a substance then we can easily work out how much of it corresponds to one mole, and by weighing this amount, and then dividing the result by Avogadro's number, we can obtain an estimate of the molecular weight).

The experimental molar heat capacities Cp at constant pressure for various solids. The heat capacity at constant volume is somewhat less than the constant pressure value, but not by much, because solids are fairly incompressible. It can be seen that Dulong and Petite's law (*i.e.*, that all solids have a molar heat capacities close to 24.9 joules/mole/degree) holds pretty well for metals. However, the law fails badly for diamond. This is not surprising. As is well-known, diamond is an extremely hard substance, so its intermolecular bonds must be very strong, suggesting that the force constant k is large. Diamond is also a fairly low density substance, so the mass m involved in lattice vibrations is comparatively small. Both these facts suggest that the typical lattice vibration frequency of diamond ($\omega \sim \sqrt{k/m}$) is high. In fact, the spacing between the different vibration energy levels (which scales like $\hbar\omega$) is sufficiently large in diamond for the vibrational degrees of freedom to be largely frozen out at room temperature.

This accounts for the anomalously low heat capacity of diamond.

Table. Values of (Joules/Mole/Degree)
For Some Solids at T = 298° K. K. From Reif.

Solid	*Cp*	Solid	*Cp*
Copper	24.5	Aluminium	24.4
Silver	25.5	Tin (white)	26.4
Lead	26.4	Sulphur (rhombic)	22.4
Zinc	25.4	Carbon (diamond)	6.1

Dulong and Petite's law is essentially a high temperature limit. The molar heat capacity cannot remain a constant as the temperature approaches absolute zero, since, by equation this would imply S ® ¥, which violates the third law of thermodynamics. We can make a crude model of the behaviour of C_V at low temperatures by assuming that all the normal modes oscillate at the same frequency, w, say.

According to equation the solid acts like a set of $3N$ independent oscillators which, making use of Einstein's approximation, all vibrate at the same frequency. We can use the quantum mechanical result for a single oscillator to write the mean energy of the solid in the form,

$$\bar{E} = 3N\hbar\omega\left(\frac{1}{2} + \frac{1}{\exp(\beta\hbar\omega) - 1}\right).$$

The molar heat capacity is defined,

$$c_V = \frac{1}{v}\left(\frac{\partial \bar{E}}{\partial T}\right)_V = \frac{1}{v}\left(\frac{\partial \bar{E}}{\partial T}\right)_V \frac{\partial \beta}{\partial T} = -\frac{1}{vkT^2}\left(\frac{\partial \bar{E}}{\partial T}\right)_V,$$

giving,

$$c_V = -\frac{3\,N_A\hbar\omega}{kT^2}\left[-\frac{\exp(\beta\hbar\omega)}{[\exp(\beta\hbar\omega) - 1]^2}\right],$$

which reduces to,

$$c_V = 3R\left(\frac{\theta_E}{T}\right)^2 \frac{\exp(\theta_E / T)}{[\exp(\theta_E / T) - 1]^2}.$$

Here,

$$\theta_E = \frac{\hbar\omega}{k}$$

is called the Einstein temperature. If the temperature is sufficiently high that T>> q_E then $kT >> \hbar\omega$, and the above expression reduces to $C_V = 3R$,, after expansion of the exponential functions. Thus, the law of Dulong and Petite is recovered for temperatures significantly in excess of the Einstein temperature. On the other hand, if the temperature is sufficiently low that T<< q_E then the exponential factors in equation become very much larger than unity, giving,

$$C_V \sim 3R\left(\frac{\theta_E}{T}\right)\exp(-\theta_E / T).$$

So, in this simple model the specific heat approaches zero exponentially as $T \to 0$.

In reality, the specific heats of solids do not approach zero quite as quickly as suggested by Einstein's model when $T \to 0$. The experimentally observed low temperature behaviour is more like C_V μ T^3. The reason for this discrepancy is the crude approximation that all normal modes have the same frequency. In fact, long wavelength modes have lower frequencies than short wavelength modes, so the former are much harder to freeze out than the latter (because the spacing between quantum energy levels, $\hbar\omega$, is smaller in the former case).

The molar heat capacity does not decrease with temperature as rapidly as suggested by Einstein's model because these long wavelength modes are able to make a significant contribution to the heat capacity even at very low temperatures.

A more realistic model of lattice vibrations was developed by the Dutch physicist Peter Debye in 1912. In the Debye model, the frequencies of the normal modes of vibration are estimated by treating the solid as an isotropic continuous medium. This approach is reasonable because the only modes which really matter at low temperatures are the long wavelength modes: *i.e.*, those whose wavelengths greatly exceed the interatomic spacing. It is plausible that these modes are not particularly sensitive to the discrete nature of the solid: *i.e.*, the fact that it is made up of atoms rather than being continuous.

Consider a sound wave propagating through an isotropic continuous medium. The disturbance varies with position vector r and time t like exp[$-i$ (k.r – wt)], where the wave-vector k and the frequency of oscillation w satisfy the dispersion relation for sound waves in an isotropic medium:

$$\omega = kC_s.$$

Here, C_S is the speed of sound in the medium. Suppose, for the sake of argument, that the medium is periodic in the x–, y–, and z–directions with periodicity lengths L_x, L_y, and L_z, respectively. In order to maintain periodicity we need

$$k_x(x + L_x) = k_x + 2\pi n_x,$$

where n_x is an integer. There are analogous constraints on k_y and k_z. It follows that in a periodic medium the components of the wave-vector are quantized, and can only take the values

$$k_x = \frac{2\pi}{L_x} n_x,$$

$$k_y = \frac{2\pi}{L_y} n_y,$$

$$k_z = \frac{2\pi}{L_z} n_z,$$

where n_x, n_y, and n_z are all integers. It is assumed that L_x, L_y, and L_z are macroscopic lengths, so the allowed values of the components of the wave-vector are very closely spaced. For given values of k_y and k_z, the number of allowed values of k_x which lie in the range k_x to $k_x + dk_x$ is given by,

$$\Delta n_x = \frac{L_x}{2\pi} 2k_x.$$

It follows that the number of allowed values of k (*i.e.*, the number of allowed modes) when k_x lies in the range k_z to $k_z + dk_z$, k_z, $+ k_y$ lies in the range k_y to $k_y + dk_y$, and k_z lies in the range k_z to $k_z + dk_z$, is,

$$\rho d^3k = \left(\frac{L_x}{2\pi} dk_x\right)\left(\frac{L_y}{2\pi} dk_y\right)\left(\frac{L_z}{2\pi} dk_z\right) = \frac{V}{(2\pi)^3} dk_x dk_y dk_z,$$

where $V = L_x L_y L_z$ is the periodicity volume, and d^3k º $dk_x dk_y dk_z$. The quantity r is called the density of modes. Note that this density is independent of k, and proportional to the periodicity volume.

Thus, the density of modes per unit volume is a constant independent of the magnitude or shape of the periodicity volume. The density of modes per unit volume when the magnitude of k lies in the range k to $k + dk$ is given by multiplying the density of modes per unit volume by the "volume" in k-space of the spherical shell lying between radii k and $k + dk$. Thus,

$$\rho_k dk = \frac{4\pi k^2 \, dk}{(2\pi)} = \frac{V^2}{2\pi^2} dk.$$

Consider an isotropic continuous medium of volume V. According to the above relation, the number of normal modes whose frequencies lie between w and w + dw (which is equivalent to the number of modes whose k values lie in the range w/C_s to w/C_s + dw/C_s) is,

$$\sigma_C(\omega)\, d\omega = 3\frac{k^2 V}{2\pi^2} dk = 3\frac{V^2}{2\pi^2 C_S{}^3}\omega^2 d\omega.$$

The factor of 3 comes from the three possible polarizations of sound waves in solids. For every allowed wavenumber (or frequency) there are two independent torsional modes, where the displacement is perpendicular to the direction of propagation, and one longitudinal mode, where the displacement is parallel to the direction of propagation. Torsion waves are vaguely analogous to electromagnetic waves (these also have two independent polarizations). The longitudinal mode is very similar to the compressional sound wave in gases. Of course, torsion waves can not propagate in gases because gases have no resistance to deformation without change of volume.

The Debye approach consists in approximating the actual density of normal modes s(w) by the density in a continuous medium s_C(w), not only at low frequencies (long wavelengths) where these should be nearly the same, but also at higher frequencies where they may differ substantially.

Suppose that we are dealing with a solid consisting of N atoms. We know that there are only 3N independent normal modes. It follows that we must cut off the density of states above some critical frequency, w_D say, otherwise we will have too many modes. Thus, in the Debye approximation the density of normal modes takes the form,

$$\sigma_D(\omega) = \sigma_C(\omega) \text{ for } \omega < \omega_D$$

$$\sigma_D(\omega) = 0 \text{ for } \omega < \omega_D$$

Here, w_D is the Debye frequency. This critical frequency is chosen such that the total number of normal modes is 3N, so,

$$\int_0^\infty \sigma_C(\omega)\, d\omega = \int_0^{\omega_D} \sigma_C(\omega)\, d\omega = 3N.$$

Substituting equation into the previous formula yields,

$$\frac{3V}{2\pi^2 C_s^3}\int_0^{\omega_D} d\omega = \frac{3V}{2\pi^2 C_s^3}\omega_D^3 = 3N.$$

This implies that,

$$\omega_D = C_s\left(6\pi^2\frac{N}{V}\right)^{1/3}.$$

Thus, the Debye frequency depends only on the sound velocity in the solid and the number of atoms per unit volume. The wavelength corresponding to the Debye frequency is 2p C_s/w_D, which is clearly on the order of the interatomic spacing $a \sim (V/N)^{1/3}$. It follows that the cut-off of normal modes whose frequencies exceed the Debye frequency is equivalent to a cut-off of normal modes whose wavelengths are less than the interatomic spacing. Of course, it makes physical sense that such modes should be absent.

Compares the actual density of normal modes in diamond with the density predicted by Debye theory. Not surprisingly, there is not a particularly strong resemblance between these two curves, since Debye theory is highly idealized. Nevertheless, both curves exhibit sharp cut-offs at high frequencies, and coincide at low frequencies. Furthermore, the areas under both curves are the same. This is sufficient to allow Debye theory to correctly account for the temperature variation of the specific heat of solids at low tem.peratures.

We can use the quantum mechanical expression for the mean energy of a single oscillator, equation to calculate the mean energy of lattice vibrations in the Debye approximation. We obtain

$$\bar{E} = \int_0^\infty \sigma_D(\omega)\hbar\omega\left(\frac{1}{2} + \frac{1}{\exp(\beta\hbar\omega) - 1}\right)d\omega.$$

According to equation the molar heat capacity takes the form,

$$C_V = \frac{1}{vkT^2}\int_0^\infty \sigma_D(\omega)\hbar\omega\left[\frac{\exp(\beta\hbar\omega)\hbar\omega}{[\exp(\beta\hbar\omega) - 1]^2}\right]d\omega.$$

Substituting in equation we find that,

$$C_V = \frac{k}{v}\int_0^{\omega_D} \frac{\exp(\beta\hbar\omega)(\beta\hbar\omega)^2}{[\exp(\beta\hbar\omega) - 1]^2}\frac{3V}{2\pi^2 C_s^3}\omega^2 d\omega,$$

giving,

$$C_V = \frac{3Vk}{2\pi^2 v(C_s\beta\hbar)^3}\int_0^{\beta\hbar\omega_D} \frac{\exp x}{(\exp x - 1)^2}x^4 dx,$$

in terms of the dimensionless variable $x = b\hbar\omega$. According to equation the volume can be written,

$$V = 6\pi^2 N\left(\frac{Cs}{\omega_D}\right)^3,$$

so the heat capacity reduces to,

$$C_V = 3Rf_D(\beta\hbar\omega_D) = 3Rf_D(\theta_D / T),$$

where the Debye function is defined,

$$f_D(y) \equiv \frac{3}{y^3}\int_0^y \frac{\exp x}{(\exp x - 1)^2}x^4 dx.$$

We have also defined the Debye temperature q_D as,

$$k\theta_D = \hbar\omega_D.$$

Consider the asymptotic limit in which $T >> q_D$. For small y, we can approximate exp x as $1 + x$ in the integrand of equation so that,

$$f_D(y) \to \frac{3}{y^3}\int_0^y x^2 dx = 1.$$

Thus, if the temperature greatly exceeds the Debye temperature we recover the law of Dulong and Petite that $C_V = 3R$. Consider, now, the asymptotic limit in which $T << q_D$. For large y,

$$\int_0^y \frac{\exp x}{(\exp x - 1)^2} x^4 dx \simeq \int_0^\infty \frac{\exp x}{(\exp x - 1)^2} x^4 dx = \frac{4\pi^4}{15}.$$

Thus, in the low temperature limit,

$$f_D(y) \to \frac{4\pi^4}{5}\frac{1}{y^3}.$$

This yields,

$$c_V \simeq \frac{12\pi^4}{5} R \left(\frac{T}{\theta_D}\right)^3$$

in the limit $T << q_D$: *i.e.*, c_V varies with temperature like T^3.

Table. Comparison of Debye Temperatures (In Degrees Kelvin)

Solid	θ_D from low temp.	θ_D from sound speed
NaCl	308	320
KCl	230	246
Ag	225	216
Zn	308	305

The fact that c_V goes like T^3 at low temperatures is quite well verified experimentally, although it is sometimes necessary to go to temperatures as low as 0.02 q_D to obtain this asymptotic behaviour.

Theoretically, q_D should be calculable from equation in terms of the sound speed in the solid and the molar volume. A comparison of Debye temperatures evaluated by this means with temperatures obtained empirically by fitting the law to the low temperature variation of the heat capacity. It can be seen that there is fairly good agreement between the theoretical and empirical Debye temperatures. This suggests that the Debye theory affords a good, thought not perfect, representation of the behaviour of c_V in solids over the entire temperature range.

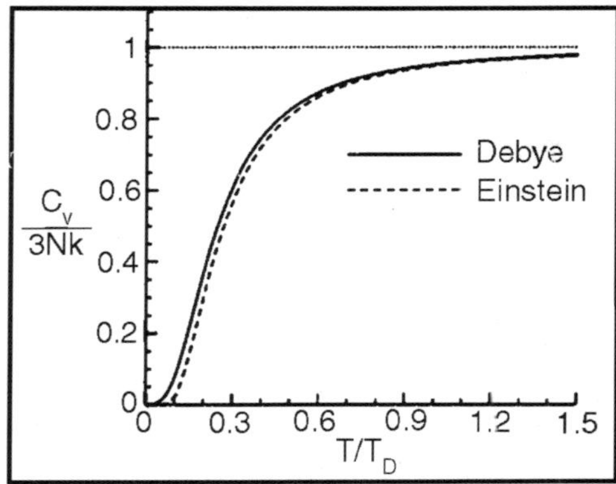

Fig. The Molar Heat Capacity of Various Solids.

Finally, the actual temperature variation of the molar heat capacities of various solids as well as that predicted by Debye's theory. The prediction of Einstein's theory is also show for the sake of comparison.

SPECIFIC HEATS OF GASES

Let us now investigate the specific heats of gases. Consider, first of all, translational degrees of freedom. Every molecule in a gas is free to move in three dimensions. If one particular molecule has mass m and momentum $p = mv$ then its kinetic energy of translation is,

$$K = \frac{1}{2m}\left(p_x^2 + p_y^2 + p_z^2\right).$$

The kinetic energy of other molecules does not involve the momentum P of this particular molecule. Moreover, the potential energy of interaction between molecules depends only on their position coordinates, and, thus, certainly does not involve P. Any internal rotational, vibrational, electronic, or nuclear degrees of freedom of the molecule also do not involve P.

Hence, the essential conditions of the equipartition theorem are satisfied (at least, in the classical approximation). Since equation contains three independent quadratic terms, there are clearly three degrees of freedom associated with translation (one for each dimension of space), so the translational contribution to the molar heat capacity of gases is,

$$(c_V)_{\text{translation}} = \frac{3}{2}R.$$

Suppose that our gas is contained in a cubic enclosure of dimensions L. According to Schrödinger's equation, the quantized translational energy levels of an individual molecule are given by,

$$E = \frac{\hbar^2\pi^2}{2mL^2}\left(n_1^2 + n_2^2 + n_3^2\right),$$

where n_1, n_2, and n_3 are positive integer quantum numbers. Clearly, the spacing between the energy levels can be made arbitrarily small by increasing the size of the enclosure. This implies that translational degrees of freedom can be treated classically, so that equation is always valid (except very close to absolute zero). We conclude that all gases possess a minimum molar heat capacity of (3/2) R due to the translational degrees of freedom of their constituent molecules.

The electronic degrees of freedom of gas molecules (*i.e.*, the possible configurations of electrons orbiting the atomic nuclei) typically give rise to absorption and emission in the ultraviolet or visible regions of the spectrum. It follows from electronic degrees of freedom are frozen out at room temperature.

Similarly, nuclear degrees of freedom (*i.e.*, the possible configurations of protons and neutrons in the atomic nuclei) are frozen out because they are associated with absorption and emission in the X–ray and g–ray regions of the electromagnetic spectrum. In fact, the only additional degrees of freedom we need worry about for gases are rotational and vibrational degrees of freedom. These typically give rise to absorption lines in the infrared region of the spectrum.

The rotational kinetic energy of a molecule tumbling in space can be written,

$$K = \frac{1}{2} I_x \omega_x^2 + \frac{1}{2} I_y \omega_y^2 + \frac{1}{2} I_z \omega_z^2,$$

where the x –, y–, and z–axes are the so called principle axes of inertia of the molecule (these are mutually perpendicular), w_x, w_y, and w_z are the angular velocities of rotation about these axes, and I_x, I_y, and I_z are the moments of inertia of the molecule about these axes. No other degrees of freedom depend on the angular velocities of rotation.

Since the kinetic energy of rotation is the sum of three quadratic terms, the rotational contribution to the molar heat capacity of gases is,

$$(c_V)_{\text{rotation}} = \frac{3}{2} R,$$

according to the equipartition theorem. Note that the typical magnitude of a molecular moment of inertia is md^2, where m is the molecular mass, and d is the typical interatomic spacing in the molecule. A special case arises if the molecule is linear (*e.g.* if the molecule is diatomic).

In this case, one of the principle axes lies along the line of centers of the atoms. The moment of inertia about this axis is of order ma^2, where a is a typical nuclear dimension (remember that nearly all of the mass of an atom resides in the nucleus). Since $a \sim 10^{-5}\, d$,it follows that the moment of inertia about the line of centres is minuscule compared to the moments of inertia about the other two principle axes. In quantum mechanics, angular momentum is quantized in units of $\hbar$. The energy levels of a rigid rotator are written,

$$E = \frac{\hbar^2}{2I} J(J+1),$$

where I is the moment of inertia and J is an integer. Note the inverse dependence of the spacing between energy levels on the moment of inertia.

It is clear that for the case of a linear molecule, the rotational degree of freedom associated with spinning along the line of centres of the atoms is frozen out at room temperature, given the very small moment of inertia along this axis, and, hence, the very widely spaced rotational energy levels.

Classically, the vibrational degrees of freedom of a molecule are studied by standard normal mode analysis of the molecular structure. Each normal mode behaves like an independent harmonic oscillator, and, therefore, contributes R to the molar specific heat of the gas [(1/2) R from the kinetic energy of vibration and (1/2)R from the potential energy of vibration].

A molecule containing n atoms has $n - 1$ normal modes of vibration. For instance, a diatomic molecule has just one normal mode (corresponding to periodic stretching of the bond between the two atoms). Thus, the classical contribution to the specific heat from vibrational degrees of freedom is,

$$(c_V)_{\text{vibration}} = (n-1)R.$$

The rotational and vibrational degrees of freedom actually make a contribution to the specific heats of gases at room temperature, once quantum effects are taken into consideration? We can answer this question by examining just one piece of data. The infrared absorption spectrum of Hydrogen Chloride. The absorption lines correspond to simultaneous transitions between different vibrational and rotational energy levels.

Hence, this is usually called a vibration-rotation spectrum. The missing line at about 3.47 microns corresponds to a pure vibrational transition from the ground-state to the first excited state (pure vibrational transitions are forbidden: HCl molecules always have to simultaneously change their rotational energy level if they are to couple effectively to electromagnetic radiation).

The longer wavelength absorption lines correspond to vibrational transitions in which there is a simultaneous decrease in the rotational energy level. Likewise, the shorter wavelength absorption lines correspond to vibrational transitions in which there is a simultaneous increase in the rotational energy level.

It is clear that the rotational energy levels are more closely spaced than the vibrational energy levels. The pure vibrational transition gives rise to absorption at about 3.47 microns, which corresponds to infrared radiation of frequency 8.5×10^{11} hertz with an associated radiation "temperature" of 4400 degrees kelvin.

We conclude that the vibrational degrees of freedom of HCl, or any other small molecule, are frozen out at room temperature. The rotational transitions split the vibrational lines by about 0.2 microns. This implies that pure rotational transitions would be associated with infrared radiation of frequency 5×10^{12} hertz and corresponding radiation "temperature" 260 degrees kelvin. We conclude that the rotational degrees of freedom of HCl, or any other small molecule, are not frozen out at room temperature, and probably contribute the classical (1/2)R to the molar specific heat. There is one proviso, however. Linear molecules (like HCl) effectively only have two rotational degrees of freedom (instead of the usual three), because of the very small moment of inertia of such molecules along the line of centres of the atoms.

We are now in a position to make some predictions regarding the specific heats of various gases. Monatomic molecules only possess three translational degrees of freedom, so monatomic gases should have a molar heat capacity (3/2)R = 12.47 joules/degree/mole.

The ratio of specific heats g = $Cp/C_V = (C_V + R)/C_V$ should be 5/3 = 1.667. It can be seen from both of these predictions are borne out pretty well for Helium and Argon. Diatomic molecules possess three translational degrees of freedom and two rotational degrees of freedom (all other degrees of freedom are frozen out at room temperature). Thus, diatomic gases should have a molar heat capacity (5/2)R = 20.8 joules/degree/mole.

The ratio of specific heats should be 7/5 = 1.4 It can be seen from these are pretty accurate predictions for Nitrogen and Oxygen. The freezing out of vibrational degrees of freedom becomes gradually less effective as molecules become heavier and more complex.

This is partly because such molecules are generally less stable, so the force constant k is reduced, and partly because the molecular mass is increased. Both these effect reduce the frequency of vibration of the molecular normal modes and, hence, the spacing between vibrational energy levels.

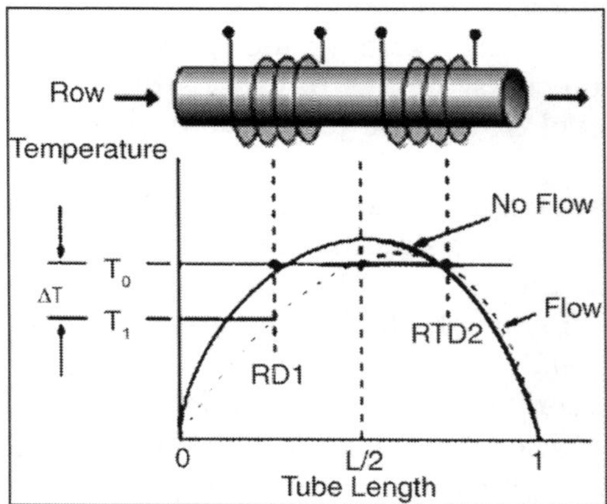

Fig. The Molar Heat Capacity at Constant Volume.

This accounts for the obviously non-classical [*i.e.,* not a multiple of (1/2)R] specific heats of Carbon Dioxide and Ethane. In both molecules, vibrational degrees of freedom contribute to the molar specific heat (but not the full R because the temperature is not high enough).

The variation of the molar heat capacity at constant volume (in units of R) of gaseous hydrogen with temperature. The expected contribution from the translational degrees of freedom is $(3/2)R$ (there are three translational degrees of freedom per molecule).

The expected contribution at high temperatures from the rotational degrees of freedom is R (there are effectively two rotational degrees of freedom per molecule). Finally, the expected contribution at high temperatures from the vibrational degrees of freedom is R (there is one vibrational degree of freedom per molecule). It can be seen that as the temperature rises the rotational, and then the vibrational, degrees of freedom eventually make their full classical contributions to the heat capacity.

HEAT CAPACITY OR SPECIFIC HEAT

Suppose that a body absorbs an amount of heat ΔQ and its temperature consequently rises by ΔT. The usual definition of the heat capacity, or specific heat, of the body is

$$C = \frac{\Delta Q}{\Delta T}.$$

If the body consists of v moles of some substance then the molar specific heat (i.e., the specific heat of one mole of this substance) is defined

$$c = \frac{1}{v}\frac{\Delta Q}{\Delta T}.$$

In writing the above expressions, we have tacitly assumed that the specific heat of a body is independent of its temperature. In general, this is not true. We can overcome this problem by only allowing the body in question to absorb a very small amount of heat, so that its temperature only rises slightly, and its specific heat remains approximately constant. In the limit as the amount of absorbed heat becomes infinitesimal, we obtain

$$c = \frac{1}{v}\frac{đW}{dT}.$$

In classical thermodynamics, it is usual to define two specific heats. Firstly, the molar specific heat at constant volume, denoted

$$c_p = \frac{1}{v}\left(\frac{đQ}{dT}\right)_V.$$

and, secondly, the molar specific heat at constant pressure, denoted

$$c_p = \frac{1}{v}\left(\frac{đQ}{dT}\right)_p.$$

Consider the molar specific heat at constant volume of an ideal gas. Since $dV = 0$, no work is done by the gas, and the first law of thermodynamics reduces to

$$đQ = dE.$$

It follows from previous equation that

$$c_V = \frac{1}{v}\left(\frac{\partial E}{\partial T}\right)_V.$$

Now, for an ideal gas the internal energy is volume independent. Thus, the above expression implies that the specific heat at constant volume is also volume independent. Since E is a function only of T, we can write

$$dE = \left(\frac{\partial E}{\partial T}\right)_V dT.$$

The previous two expressions can be combined to give

$$dE = v\, c_V\, dT$$

for an ideal gas.

Let us now consider the molar specific heat at constant pressure of an ideal gas. In general, if the pressure is kept constant then the volume changes, and so the gas does work on its environment. According to the first law of thermodynamics,

$$đQ = dE + pdV = vc_V\, dT + pdV.$$

The equation of state of an ideal gas tells us that if the volume changes by dV, the temperature changes by dT, and the pressure remains constant, then

$$pdV = vRdT.$$

The previous two equations can be combined to give

$$đQ = vVdT + vRdT.$$

Now, by definition

$$c_p = \frac{1}{v}\left(\frac{đQ}{dT}\right)_p,$$

so we obtain

$$c_p = c_V + R.$$

for an ideal gas. This is a very famous result. Note that at constant volume all of the heat absorbed by the gas goes into increasing its internal energy, and, hence, its temperature, whereas at constant pressure some of the absorbed heat is used to do work on the environment as the volume increases. This means that, in the latter case, less heat is available to increase the temperature of the gas. Thus, we expect the specific heat at constant pressure to exceed that at constant volume, as indicated by the above formula.

The ratio of the two specific heats Cp/C_V is conventionally denoted γ. We have

$$\gamma \equiv \frac{c_p}{c_V} = 1 + \frac{R}{c_V}.$$

for an ideal gas. In fact, γ is very easy to measure because the speed of sound in an ideal gas is written

$$c_s = \sqrt{\frac{\gamma p}{\rho}},$$

where $\rho$is the density. The extent of the agreement between g calculated from previous equation and the experimental g is quite remarkable.

Table. Specific Heats of Common Gases in Joules/mole/deg.
(at 15 °C and 1 atm.)

Gas	Symbol	c_V	γ	γ
		(experiment)	(experiment)	(theory)
Helium	He	12.5	1.666	1.666
Argon	Ar	12.5	1.666	1.666
Nitrogen	N_2	20.6	1.405	1.407
Oxygen	O_2	21.1	1.396	1.397
Carbon Dioxide	CO_2	28.2	1.302	1.298
Ethane		39.3	1.22	1.214

CALCULATION OF SPECIFIC HEATS

Now that we know the relationship between the specific heats at constant volume and constant pressure for an ideal gas, it would be nice if we could calculate either one of these quantities from first principles. Classical thermodynamics cannot help us here. However, it is quite easy to calculate the specific heat at constant volume using our knowledge of statistical physics. Recall, that the variation of the number of accessible states of an ideal gas with energy and volume is written

$$\Omega(E,V) \propto V^N \chi(E).$$

For the specific case of a monatomic ideal gas, we worked out a more exact expression for Ω,

$$\Omega(E,V) = B\, V^N E^{3N/2},$$

where B is a constant independent of the energy and volume. It follows that

$$\ln\Omega = \ln B + N\ln V + \frac{3N}{2}\ln E.$$

The temperature is given by

$$\frac{1}{kT} = \frac{\partial \ln\Omega}{\partial E} = \frac{3N}{2}\frac{1}{E},$$

so

$$E = \frac{3}{2}N\,kT.$$

Since, $N = v\,N_A$, and $N_A\,k = R$, we can rewrite the above expression as

$$E = \frac{3}{2}v\,RT,$$

where R = 8.3143 joules/mole/deg, is the ideal gas constant. The above formula tells us exactly how the internal energy of a monatomic ideal gas depends on its temperature.

The molar specific heat at constant volume of a monatomic ideal gas is clearly

$$c_V = \frac{1}{V}\left(\frac{\partial E}{\partial T}\right)_v = \frac{3}{2}R.$$

This has the numerical value

$$c_V = 12.47 \text{ joules/mole/deg.}$$

Furthermore, we have

$$cp = cV + R = \frac{5}{2}R,$$

and

$$\gamma \equiv \frac{c_p}{c_V} = \frac{5}{3} = 1.667.$$

The predictions are borne out pretty well for the monatomic gases Helium and Argon. Note that the specific heats of polyatomic gases are larger than those of monatomic gases. This is because polyatomic molecules can rotate around their centres of mass, as well as translate, so polyatomic gases can store energy in the rotational, as well as the translational, energy states of their constituent particles.

11

Enthalpy and Specific Heat Capacities of Ideal Gas

DEFINITION OF GAS

Gases are material of very low density that can be enclosed to keep together. Gases have no definite shape and volume, but they can completely fill a container. The volume of the container is the volume of the gas in it. A gas exerts a pressure on all sides of the container that holds it. Vapour, fume, air, or miasma also describes a gas. Air describes the common mixture of gases in the atmosphere. Liquid air has a density of 0.875 grams per milliliter. By Avogadro's law, a mol of any gas occupies 22.4 liters at standard temperature and pressure (STP).

1 mol of any gas at STP = 22.4 liters

When a gas molecule hits one another, they try to bounce off in a completely elastic encounter. There is pressure within the gas that is caused by the gas molecules in motion striking each other and anything else in the gas. The pressure that a gas exerts on its container that comes from the molecules of gas hitting the inside of the container and bouncing off.

TYPES OF GASES

There are different types of gases that are categorized as bulk gases as: Argon, Carbon Dioxide, Hydrogen, Helium, Oxygen, Nitrogen, Nitrous Oxide, Propane and Propylene.

Bulk Carbon Dioxide

Production: Raw carbon dioxide is purified in many stages. It is compressed and cooled to obtain liquid carbon dioxide at a pressure of 20 bg and at a temperature of 20 degrees. It is collected and stored in large tanks.

Quality: Bulk Liquefied carbon dioxide, which is supplied to the beverage and food industries, has the most stringent and detailed product specifications.

Gaseous properties: Colourless, tasteless, odourless, dry and soluble in water.

Liquid properties: At atmospheric pressure, it cannot exist. It becomes super critical when it is pumped and heated above its critical point.

Bulk Liquid Nitrogen

Nitrogen (N_2) is the major gas in our atmosphere; making up about 78 per cent of it. The commercially available N_2 is made from bigger 'cryogenic air separation plants' and in this type of plant atmospheric air is completely cleaned and the gas is compressed and later it is cooled by expansion to create the liquid gas. This liquid gas is stored or transported at a temperature ranging from -196 to155 degrees Celsius.

Gaseous properties: Colourless, tasteless and odourless. Non-flammable at very high temperatures. Slightly soluble in water.

Liquid properties: When it vaporizes, it absorbs a lot of heat. Has very high heat transfer rate.

Bulk Liquid Oxygen

Oxygen (O_2) is one of the important gases the atmosphere, and it comprises about 21 per cent of air. The commercially available O_2 is made from bigger 'cryogenic air separation plants' and in this type of plant, atmospheric air is completely cleaned and the gas is compressed and later it is cooled by expansion to create the liquid gas, and will be separated by fractional distillation. This liquid gas is stored or transported at a temperature ranging from -183 to 143 degrees Celsius.

Gaseous properties: Colourless, tasteless and odourless. Non flammable, but supports and enhances combustion of other materials. Liquid properties: Blue in colour.

Bulk Liquid Argon

This gas comprises just 1 per cent of the earth's atmosphere. The commercially available argon is made from bigger 'cryogenic air separation plants' and in this type of plant atmospheric air is completely cleaned and the gas is compressed and later it is cooled by expansion to create the liquid gas and will be separated by fractional distillations. Industrial based argon contains slight amounts of oxygen, and if pure argon is wanted, the gas needs more purification and distillation. Liquefied argon should be stored at temperatures ranging from -186 to -143 ºC and also the pressure requirement should be about 20 bar.

- *Gaseous properties:*
- Colourless
- Tasteless
- Odourless

- Soluble in Water
- Non flammable
- Extremely inert

PERFECT GAS LAW

A gas may be completely described by its pressure, temperature, and volume.

The "Ideal Gas Law Formula" can be written as:

P V = n R T

Where:

P is the pressure

V is the volume

n is the number of moles of gas

T is the absolute temperature

R is the Universal Gas Constant,

The Universal Gas Constant, R, can be expressed in several ways, depending upon the units of P, V, and T. One common R is 0.0821 liter - atmospheres per mol - degree.

The pressure of a fixed volume of all gases is found to vary linearly with temperature according to,

$$Pt = Po\ (1 + A\ t)$$

Here Pt is the pressure at temperature t, and Po is the pressure at the zero (t = 0) of the particular scale used. A is constant.

Look at equation (1) and notice that for t = – 1/A, Pt = 0.

At temperatures lower than this point the pressure would become negative, which is impossible. Thus this temperature is called the absolute zero. Since equation (1) is same for all gases, it can be used to define a new scale of temperature, the absolute or Kelvin scale. Call temperatures on this new scale T and set,

$$t = -1/A + T$$

Then equation (1) becomes,

$$P = Po\ A\ T$$

If the volume of the gas is now varied it is found that the complete equation may be written (for n moles of the gas).

This is the perfect gas law, where R is the universal gas constant.

ENTHALPY OF IDEAL GAS

Under conditions of *constant pressure* the heat absorbed or released is termed as *enthalpy*. We cannot measure enthalpy *directly*, but can calculate the *change* in enthalpy DH.

The change in enthalpy, DH, equals the heat, q_p, added to or lost by the system when the process occurs under constant pressure:

$$DH = q_p$$

DH represents the difference between the enthalpy of the system at the beginning of the reaction compared to what it is at the end of the reaction:

$$DH = H_{final} - H_{initial}$$

We are considering the enthalpic state of the *system*. Thus:

- If the *system* has higher enthalpy at the end of the reaction, then it *absorbed* heat from the surroundings (*endothermic* reaction)
- If the system has a lower enthalpy at the end of the reaction, then it *gave off heat* during the reaction (*exothermic* reaction)

Therefore:

- For *endothermic* reactions $H_{final} > H_{initial}$ *and DH is positive (+DH)*
- For exothermic reactions $H_{final} < H_{initial}$ *and DH is negative (–DH)*

For an ideal gas the enthalpy - *h* - is function of temperature and the change in enthalpy can be expressed as,

$$dh = c_p \, dT$$

where,

dh = change in enthalpy

c_p= *specific heat capacity for the gas in a constant pressure process*

c_p can within a moderate temperature change be regarded as constant.

The enthalpy in a fluid is defined as:

$$h = u + p / \tilde{n}$$

where,

h = enthalpy

u = internal energy

p = absolute pressure

ñ = density

Combining previous equation and the Ideal Gas Law gives:

$$h = u + R\,T$$

where,

R = the individual gas constant

The change in enthalpy can be expressed by differentiating:

$$dh = du + R\,dT$$

Dividing previous equation with dT gives:

$$(dh / dT) - (du / dT) = R$$

Modifying equations:

$$c_p - c_v = R$$

The difference $c_p - c_v$ is constant for an ideal gas.

SPECIFIC HEAT CAPACITIES OF IDEAL GAS

If we are dealing with a gas, it is most convenient to use forms of the thermodynamics equations based on the enthalpy of the gas. From the definition of enthalpy:

$$h = e + p * v$$

where h in the specific enthalpy, p is the pressure, v is the specific volume, and e is the specific internal energy. During a process, the values of these variables change. Let's denote the change by the Greek letter delta which looks like a triangle. So "delta h" means the change of "h" from state 1 to state 2 during a process.

$$\text{delta } h = \text{delta } e + p * \text{delta } v$$

The enthalpy, internal energy, and volume are all changed, but the pressure remains the same. From our derivation of enthalpy equation, the change of specific enthalpy is equal to the heat transfer for a constant pressure process:

$$\text{delta } h = cp * \text{delta } T$$

where delta T is the change of temperature of the gas during the process, and c is the specific heat capacity. We have added a subscript "p" to the specific heat capacity to remind us that this value only applies to a constant pressure process.

The equation of state of a gas relates the temperature, pressure, and volume through a gas constant R . The gas constant used by aerodynamicists is derived from the universal gas constant, but has a unique value for every gas.

$$p * v = R * T$$

If we have a constant pressure process, then:

$$p * \text{delta } v = R * \text{delta } T$$

Now let us imagine that we have a constant volume process with our gas that produces exactly the same temperature change as the constant pressure process that we have been discussing. Then the first law of thermodynamics tells us:

$$\text{delta } e = \text{delta } q - \text{delta } w$$

where q is the specific heat transfer and w is the work done by the gas. For a constant volume process, the work is equal to zero. And we can express the heat transfer as a constant times the change in temperature. This gives:

$$\text{delta } e = cv * \text{delta } T$$

where delta T is the change of temperature of the gas during the process and c is the specific heat capacity.

If we substitute the expressions for "delta e", "p * delta v", and "delta h" into the enthalpy equation we obtain:

$$cp * \text{delta } T = cv * \text{delta } T + R * \text{delta } T$$

dividing by "delta T" gives the relation:

$$cp = cv + R$$

We can define an additional variable called the specific heat ratio, which is given the Greek symbol "gamma", which is equal to cp divided by cv:

$$gamma = cp / cv$$

"Gamma" is just a number whose value depends on the state of the gas. For air, gamma = 1.4 for standard day conditions. "Gamma" appears in many fluids equations including the equation relating pressure, temperature, and volume during a simple process, the equation for the speed of sound, and all the equations for isentropic flows, and shock waves.

P – V –T SURFACE OF AN IDEAL GAS

For a fixed number of molecules the Ideal Gas Law forms the surface of a three-dimensional graph where the axis are Pressure, Volume, and Temperature.

Lines of constant pressure, constant volume, and constant temperature form a coordinate system labeling the location of an ideal gas.

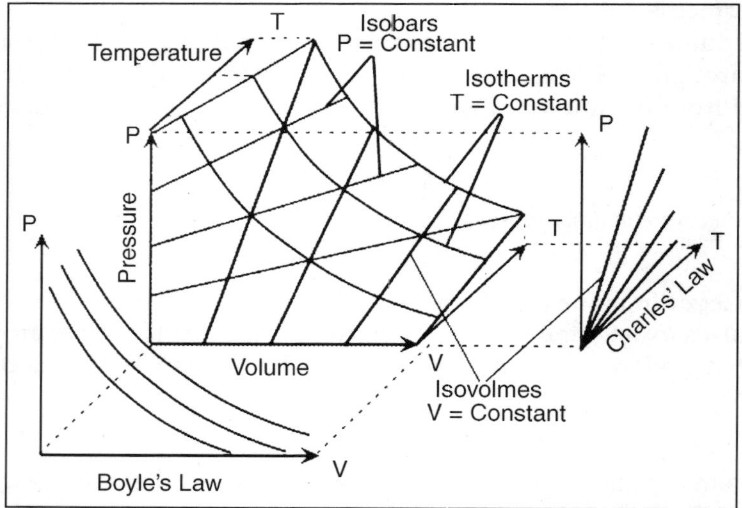

Boyle law showed that the pressure of a low-density gas is inversely proportional to its volume, when the temperature is constant.

$P \alpha_1/V$ for constant temperature.

Charles and Gay-Lussac law showed that the pressure of a low-density gas is proportional to its temperature, when the volume is constant.

$P \alpha T$ for constant volume.

The ideal gas law is only valid for low-density gas. There exists a PVT surface for high-density gases only.

PVT SURFACE FOR SUBSTANCE WHICH CONTRACTS ON FREEZING

The equilibrium state of a simple, compressible substance can be specified in terms of its pressure, volume and temperature. If any two of these state variables is specified, we can determine the third. The states of the substance can be represented as a surface in three dimensional PVT space. PVT surface above represents a substance which contracts on freezing. PVT surface image is shown in figure.

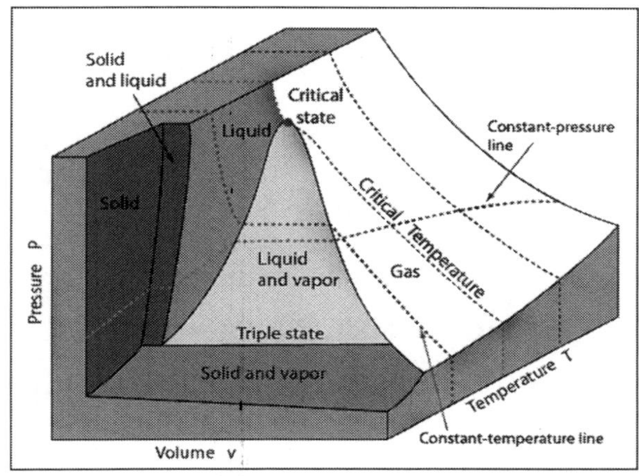

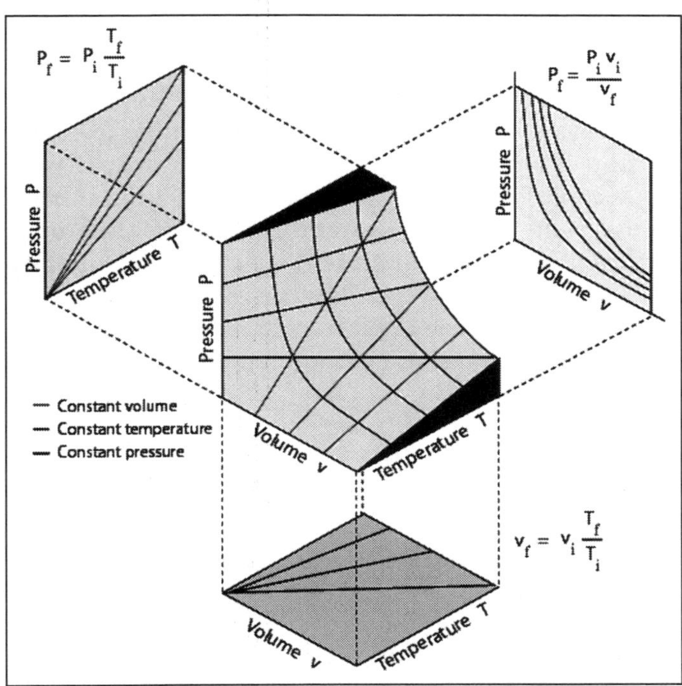

In this the solid, liquid and gas phases can be determined by regions on the surface. A point lying on a line between a single-phase and a two-phase region represents a "saturation state". The line between the liquid and liquid-vapour regions is called as liquid-saturation line. A point on the boundary between the vapour and liquid-vapour regions is called as a saturated-vapour state. All the possible states of an ideal gas can be represented by a PVT surface as shown in figure above.

TRIPLE POINT

Triple Point of a substance is defined as temperature and pressure at which three phases such as gas, liquid, and solid of that substance coexist in equilibrium.

For example, triple point of mercury occurs at a temperature of -38.8344 °C and a pressure of 0.2 mPa. In addition to the triple point between solid, liquid, and gas, there can be triple points involving more than one solid phase, for substances with multiple polymorphs. Helium-4 is a special case that presents a triple point involving two different fluid phases.

THE EQUATION OF STATE OF AN IDEAL GAS

Let us start our discussion by considering the simplest possible macroscopic system: i.e., an ideal gas. All of the thermodynamic properties of an ideal gas are summed up in its equation of state, which determines the relationship between its pressure, volume, and temperature. Unfortunately, classical thermodynamics is unable to tell us what this equation of state is from first principles.

In fact, classical thermodynamics cannot tell us anything from first principles. We always have to provide some information to begin with before classical thermodynamics can generate any new results. This initial information may come from statistical physics (i.e., from our knowledge of the microscopic structure of the system under consideration), but, more usually, it is entirely empirical in nature (i.e., it is the result of experiments).

Of course, the ideal gas law was first discovered empirically by Robert Boyle, but, nowadays, we can justify it from statistical arguments. Recall, that the number of accessible states of a monotonic ideal gas varies like

$$\Omega \propto V^N \chi(E),$$

where N is the number of atoms, and $\chi(E)$ depends only on the energy of the gas (and is independent of the volume). We obtained this result by integrating over the volume of accessible phase-space. Since the energy of an ideal gas is independent of the particle coordinates (because there are no interatomic forces), the integrals over the coordinates just reduced to N simultaneous volume integrals, giving the V^N factor in the above expression. The integrals over the particle momenta were more complicated, but were clearly completely independent of V, giving the $\chi(E)$ factor in the above expression. Now, we have a statistical rule which tells us that

$$X_a - \frac{1}{\beta}\frac{\partial \ln \Omega}{\partial x_a},$$

where X_a is the mean force conjugate to the external parameter x_a (i.e., $đW = \sum_a X_a dx_a$), and $\beta = 1/kT$. For an ideal gas, the only external parameter is the volume, and its conjugate force is the pressure (since $đW = pdV$). So, we can write

$$p = \frac{1}{\beta}\frac{\partial \ln \Omega}{\partial V}.$$

If we simply apply this rule to previous equation we obtain

$$p = \frac{N\,k\,T}{V}.$$

However, $N = v\,N_A$, where v is the number of moles, and N_A is Avagadro's number. Also, $k\,N_A = R$, where R is the ideal gas constant. This allows us to write the equation of state in its usual form

$$pV = vRT.$$

The above derivation of the ideal gas equation of state is rather elegant. It is certainly far easier to obtain the equation of state in this manner than to treat the atoms which make up the gas as little billiard balls which continually bounce of the walls of a container. The latter derivation is difficult to perform correctly because it is necessary to average over all possible directions of atomic motion. It is clear, from the above derivation, that the crucial element needed to obtain the ideal gas equation of state is the absence of interatomic forces. This automatically gives rise to a variation of the number of accessible states with E and V of the form which, in turn, implies the ideal gas law. So, the ideal gas law should also apply to polyatomic gases with no interatomic forces.

Polyatomic gases are more complicated that monatomic gases because the molecules can rotate and vibrate, giving rise to extra degrees of freedom, in addition to the translational degrees of freedom of a monatomic gas. In other words, $\chi(E)$, in earleir equation becomes a lot more complicated in polyatomic gases.

However, as long as there are no interatomic forces, the volume dependence of Ω is still V^N, and the ideal gas law should still hold true. In fact, we shall discover that the extra degrees of freedom of polyatomic gases manifest themselves by increasing the specific heat capacity. There is one other conclusion we can draw from previous equation. The statistical definition of temperature is

$$\frac{1}{kT} = \frac{\partial \ln \Omega}{\partial E}.$$

It follows that

$$\frac{1}{kT} = \frac{\partial \ln \chi}{\partial E}.$$

We can see that since χ is a function of the energy, but not the volume, then the temperature must be a function of the energy, but not the volume. We can turn this around and write

$$E = E(T).$$

In other words, the internal energy of an ideal gas depends only on the temperature of the gas, and is independent of the volume. This is pretty obvious, since if there are no interatomic forces then increasing the volume, which effectively increases the mean separation between molecules, is not going to affect the molecular energies in any way. Hence, the energy of the whole gas is unaffected.

The volume independence of the internal energy can also be obtained directly from the ideal gas equation of state. The internal energy of a gas can be considered as a general function of the temperature and volume, so

$$E = E(T, V).$$

It follows from mathematics that

$$dE = \left(\frac{\partial E}{\partial T}\right)_V dT + \left(\frac{\partial E}{\partial T}\right)_T dV,$$

where the subscript V reminds us that the first partial derivative is taken at constant volume, and the subscript T reminds us that the second partial derivative is taken at constant temperature. Thermodynamics tells us that for a quasi-static change of parameters

$$T\,dS = dE + pdV.$$

The ideal gas law can be used to express the pressure in term of the volume and the temperature in the above expression. Thus,

$$dS = \frac{1}{T}dE + \frac{vR}{V}dV.$$

Using above equation this becomes

$$dS = \frac{1}{T}\left(\frac{\partial E}{\partial T}\right)_V dT + \left[\frac{1}{T}\left(\frac{\partial E}{\partial V}\right)_T + \frac{vR}{V}\right]dV.$$

However, dS is the exact differential of a well-defined state function, S. This means that we can consider the entropy to be a function of temperature and volume. Thus, $S = S(T, V)$, and mathematics immediately tells us that

$$dS = \left(\frac{\partial E}{\partial T}\right)_V dT + \left(\frac{\partial E}{\partial V}\right)_T dV.$$

The above expression is true for all small values of dT and dV, so a comparison with given equation gives

$$\left(\frac{\partial S}{\partial T}\right)_V = \frac{1}{T}\left(\frac{\partial E}{\partial T}\right)_V,$$

$$\left(\frac{\partial S}{\partial T}\right)_T = \frac{1}{T}\left(\frac{\partial E}{\partial V}\right)_T + \frac{vR}{V}.$$

One well-known property of partial differentials is the equality of second derivatives, irrespective of the order of differentiation, so

$$\frac{\partial^2 S}{\partial V\,\partial T} = \frac{\partial^2 S}{\partial T\,\partial V}.$$

This implies that

$$\left(\frac{\partial}{\partial V}\right)_T \left(\frac{\partial S}{\partial T}\right)_V = \left(\frac{\partial}{\partial T}\right)_V \left(\frac{\partial S}{\partial T}\right)_T.$$

The above expression can be combined with Eqs. to give

$$\frac{1}{T}\left(\frac{\partial^2 E}{\partial V\,\partial T}\right) = \left[\frac{1}{T^2}\left(\frac{\partial E}{\partial V}\right)_T + \frac{1}{T}\left(\frac{\partial^2 E}{\partial T\,\partial V}\right)\right].$$

Since second derivatives are equivalent, irrespective of the order of differentiation, the above relation reduces to

$$\left(\frac{\partial E}{\partial V}\right)_T = 0,$$

which implies that the internal energy is independent of the volume for any gas obeying the ideal equation of state. This result was confirmed experimentally by James Joule in the middle of the nineteenth century.

12

Heat Conduction

INTRODUCTION

When there exists a temperature gradient within a body, heat energy will flow from the region of high temperature to the region of low temperature. This phenomenon is known as conduction heat transfer, and is described by Fourier's Law (named after the French physicist Joseph Fourier),

$$q = -k\vec{\nabla}T$$

This equation determines the heat flux vector q for a given temperature profile T andthermal conductivity k. The minus sign ensures that heat flows down the temperature gradient. The temperature profile within a body depends upon the rate of its internally-generated heat, its capacity to store some of this heat, and its rate of thermal conduction to its boundaries (where the heat is transfered to the surrounding environment). Mathematically this is stated by the Heat Equation,

$$\vec{\nabla}T - \frac{1}{\alpha}\frac{\partial T}{\partial t} = -\frac{1}{k}q\,gen$$

along with its boundary conditions, equations that prescribe either the temperature Ton, or the heat flux q through, all of the body boundaries ?,

$$T(\Omega_a) = T_{prescribed}$$

$$q(\Omega_b) = q_{prescribed}$$

$$\Omega a \cup \Omega b = \Omega$$

In the Heat Equation, the power generated per unit volume is expressed by qgen. The thermal diffusivity a is related to the thermal conductivity k, the specific heat c, and the density r by,

$$a = \frac{k}{\rho c}$$

For Steady State problems, the Heat Equation simplifies to,

$$\vec{\nabla}^2 T = -\frac{1}{k} q_{gen}.$$

DERIVATION

The heat equation follows from the conservation of energy for a small element within the body,

heat conducted in	+	heat generated within	=	heat conducted out	+	change in energy stored within

We can combine the heats conducted in and out into one "net heat conducted out" term to give,

Net heat conducted out	=	Heat generated within	–	Change in energy stored within

Mathematically, this equation is expressed as,

$$\vec{\nabla}.q = q\,gen - \frac{de}{dt}$$

The change in internal energy e is related to the body's ability to store heat by raising its temperature, given by,

$$\frac{de}{dt} = \rho c \frac{dT}{dt}$$

One can substitute for q using Fourier's Law of heat conduction from above to arrive at the Heat Equation,

$$\vec{\nabla}.\left(-k\vec{\nabla}T\right) = q_{gen} - \rho c \frac{dT}{dt}$$

$$-k\vec{\nabla}T + \rho c \frac{\partial T}{\partial t} = q_{gen}$$

$$\vec{\nabla}^2 T - \frac{1}{\alpha}\frac{\partial T}{\partial t} = -\frac{1}{k} q_{gen}$$

GENERAL HEAT CONDUCTION EQUATION

Heat or heat transfer is not directly measurable. In fact thermodynamics principally assumed the existence of heat and related the increase of the amount of heat or thermal energy dU to the increase of temperature dT through the mass m and the heat capacity at constant pressure cp ;

$$dU = mc_p dT$$

The heat capacity is determined by relating the temperature rise to the energy applied to the system at hand. The heat capacity of many substances has been determined so that eq.allows for the indirect measurement of the thermal energy contained within some volume. The change of heat content in some volume then is proportional to the change of temperature;

$$\frac{dU}{dt} = mc_p \frac{dT}{dt}$$

A change in the amount of thermal energy contained within some volume can be brought about in only two ways. Either some net exchange of heat between volume and surroundings occurred or some heat was produced within the volume. To start with the first option, the second law of thermodynamics teaches that the flow of heat is determined by the local differences in temperature. Heat flows from locations of high temperature to those of lower temperature. As mentioned this can happen through various mechanisms. The mechanism under study in this experiment is conduction and it was Fourier who described this phenomenon somewhat more quantitatively. Fourier found that, in homogenous media, heat transfer is proportional to the temperature gradient, a relation also known as Fourier's law;

$$q'' = -k\nabla T$$

Here q'' is the heat current density in [W /m^2]. The constant of proportionality is called the thermal conductivity, measured in [W / m K]. It is to be determined experimentally. The total heat transferred per unit time through some surface A than is;

$$q = Aq'' = -kA\nabla T$$

Apart from the net transport of heat a change of the thermal energy content in some volume can arise due to production of thermal energy within some volume. Of course energy can not actually be produced, but it can be transformed. Production of thermal energy then comes down to the transfer of some other form of energy into the volume, where it is transformed into thermal energy. Assuming that an electrical heating element is used, electric energy is transformed into heat. The amount of energy transformed into heat per unit time then equals the the power supplied;

$$P = VI$$

The net inflow of thermal energy and the net production of thermal energy within some volumetric element should yield the change of thermal energy.

EQUATION

In the first place a steady state temperature profile is to be generated. In the most general sense, one could choose for planar, cylindrical or spherical geometry. Both the planar and the spherical geometry are not so easily realised, since these require either a planar or a spherical heating element with well distributed heating capacity. The cylindrical geometry can be approached fairly well, using a long coil with a high accuracy in the number of windings per unit length. Therefore eq. the equation of heat transfer, is to be evaluated in cylindrical coordinates.

As mentioned before the system is evaluated for the case of a steady state, *i.e.*

$$\frac{\partial T}{\partial t} = 0 .$$

Furthermore it is assumed that the cylinder is of infinite length and therefore that heat is only transferred in the radial direction. Finally the equation is evaluated for the region around the heating rod, so that within the volume of evaluation, no production takes place. Under these conditions, eq.may be rewritten as a second order differential equation;

$$\frac{d^2T}{dr^2} + \frac{1}{r}\frac{dT}{dr} = 0 .$$

The solution to this differential equation is:

$$T = c_1 \; In \; r + c_2$$

With boundary conditions,

$$\begin{cases} T = T_i \; . \; T = T_i \\ T = T_u \; . \; T = T_u \end{cases}$$

this finally yields the radial temperature distribution;

$$T(r) = T_u + \frac{T_i - T_u}{In(r_i / r_u)} in(r / r_u)$$

In figure a sketch is presented of the temperature profile, as a function of the radial position.

For the determination of the thermal conductivity the temperature profile is to be related to the heat transfer through Fourier's law for cylindrical geometry. Since the total heat transfer through the surface should be equal to the production within the volume enclosed as given by eq., one may write;

$$P = q = Aq'' = -2\pi r LK \frac{\partial t(r)}{\partial r} = 2\pi LK \frac{T_i - T_u}{In(r_u / r_i)}$$

This relates the thermal conductivity k to the total power lost by the heating element.

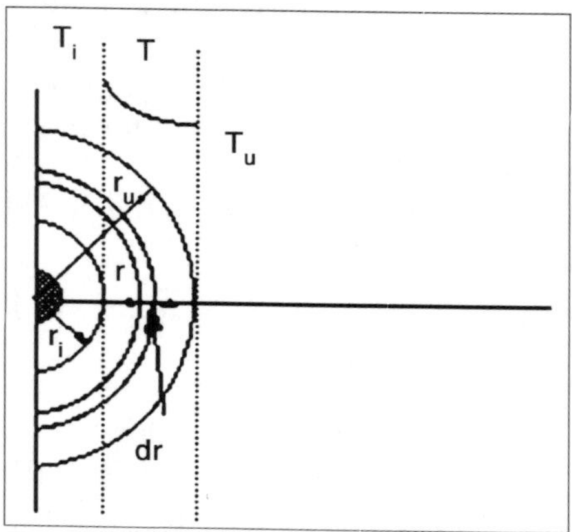

Fig. Geometry of the Cylindrical Stationary Temperature Profile.

The temperature profile is evaluated in a cylindrical volume around the heating rod. Here ri and ru respectively represent the inner and the outer radius of the region under evaluation, which will be smaller than the container of the experiment.

THE CONTROLLING COEFFICIENTS

A control coefficient is a relative measure of how much a perturbation affects a system variable (*e.g.* fluxes or concentrations). It is defined (Kacser and Burns 1973, Heinrich and Rapoport 1974, Burns *et al.* 1985) as:

$$C_i^A = \frac{\partial A}{\partial v_i} \cdot \frac{v_i}{A} = \frac{\partial \ln A}{\partial \ln v_i}$$

where *A* is the variable, *i* the step (enzyme) and *vi* its steady-state rate of the step perturbed. The most common control coefficients are those for fluxes and metabolite concentrations. However, any variable of the system can be analysed with MCA and have control coefficents defined by equations analogous to equation. In fact, there is no need even for the system to be in a steady state. Any limit set can have control coefficients defined for any of its variables which are invariant (such as the period of an oscillation; Markus and Hess 1990). They were also defined for ordinary points of a trajectory in which case there are additional time components that make the analysis more difficult.

Since the rate of reaction cannot be perturbed directly, control coefficients must be determined by perturbations in parameters that affect the rate *linearly*. Because many enzyme-catalysed reaction rates are linear in terms of the enzyme concentration (at least in a certain range of enzyme concentrations), control coefficients can be written using enzyme concentrations:

$$C_i^A = \frac{\partial A}{\partial [E_i]} = \frac{\partial \ln A}{\partial \ln [E_i]}$$

It shows example of the relation between the flux of a pathway to the concentration of one of the enzymes of that pathay. The flux-control coefficient of the enzyme is the slope of the tangent to the curve. It is easy to see that the flux-control coefficients are different from one steady state to another.

A very important property of steady-state metabolic systems was uncovered with the MCA formalism. This concerns the summation of all the flux control coefficients of a pathway. By various procedures (Kacser and Burns 1973, Heinrich and Rapoport 1975, Giersch 1988, Reder 1988) it can be demonstrated that for a given flux the sum of its flux-control coefficients of all steps in the system is equal to unity. The simplest way in which this property can be derived is by considering a simultaneous small relative increase (a) in all reaction rates of a metabolic system. Because for each metabolite the relative rates of its production increase exactly by the same amount (a) as the relative rates of its consumption, the metabolite concentrations remain unchanged. The flux of the system increases exactly by a. In mathematical terms this means that the flux is a homogeneous function of degree one and the metabolite concentrations homogeneous functions of degree zero. The summation theorems follow from this by applying a corollary of the Euler theorem for homogeneous functions (Giersch 1988). For flux-control coefficients:

$$\sum_i C_i^J = 1$$

and for concentration-control coefficients:

$$\sum_i C_i^{[S_j]} = 0$$

where the summations are over all the steps of the system. This may include not only the steps of the pathway of interest but also of other pathways (as long as there are links between them). In principle for a whole cell, the summation would have to be over all metabolic steps of that cell.

According to equation, increases in some of the flux-control coefficients imply decreases in the others so that the total remains unity. As a consequence of the summation theorems, one concludes that the control coefficients are globalproperties and that in metabolic systems,control is a systemic property, dependent on all of its elements (steps).

MAGNITUDE OF THE ANGULAR VELOCITY

A THIN ROD

Moment of inertia is independent of the magnitude of the angular velocity. Is is characteristic of the rigid body and the axis about which it rotates.

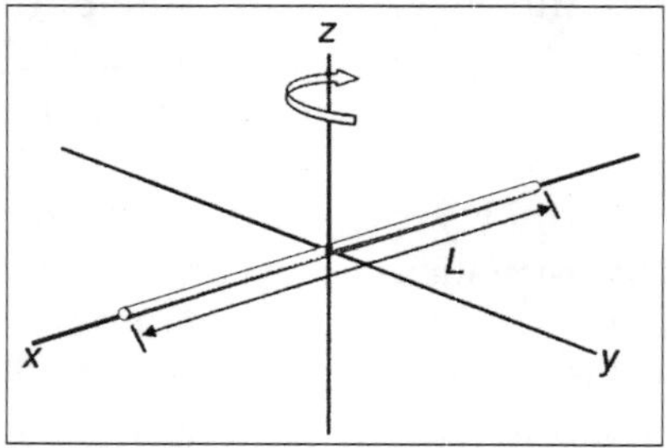

Consider a thin rod of mass M and length L which is rotating about an axis perpendicular to rod and passing through its midpoint. Now for such thin rod radius of gyration(k) is given as:

$k^2 = L^2 / 12$

$I = Mk^2$

$I = ML^2/12$

The expression for the moment of inertia of a solid cylinder can be built up from the moment of inertia of thin cylindrical shells.

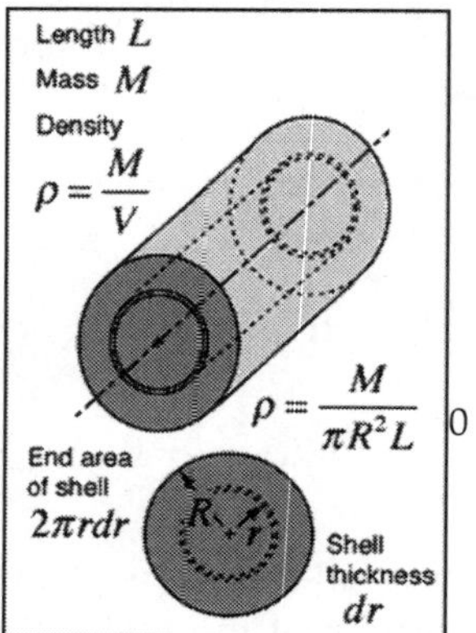

Using the general definition for moment of inertia:

$$I = \int_0^M r^2 dm$$

The mass element can be expressed in terms of an infinitesmal radial thickness dr by,

$$dm = \rho dV = \rho L 2\pi r dr$$

Substituting gives a polynomial form integral:

$$I = 2\pi\rho L \int_0^R r^3 dr = 2\pi\rho L \frac{R^4}{4}$$

$$I = 2\pi \left[\frac{M}{\pi R^2 L} \right] L \frac{R^4}{4} = \frac{1}{2} MR^2$$

CONSIDERING THE ONE DIMENSIONAL FIN

THE RECTANGULAR FIN

Considering the one dimensional fin exposed to a surrounding fluid at a temperature T. The temperature of the base of the fin is T.

One dimensional conduction and convection through a rectangular fin.

Be making an energy balance on an element of the fin of thickness dx

Energy in left face = Energy at right face +Energy lost by convection

Energy in left face,

Energy out at right face, = –kA |x+dx

Energy lost by convection = hPdx

Here,

A = cross sectional area of fin

P= perimeter of fin

The energy balance yields,

Let

Then the equation becomes,

One boundary condition is

At x=0

The other boundary condition depends on physical situation. Several cases may be considered.

Case 1: The fin is very long and the temperature at the end of the fin is essentially that of the surrounding fluid.

Case 2: The fin is of finite length and loses heat by convection from its end.

Case 3: The end of fin is insulated so that dT/dx == 0 at x =L

Considering Case 2:

Temperature distribution along the fin,

Where,

At base, X=0, Now, heat transfer rate at X = 0,

FIN EFFICIENCY AND EFFECTIVENESS

The fin efficiency and fin effectiveness are related to the performance of the fin, but they are different quantities. However, they are related to each other by,

$$\varepsilon_{fin} = \frac{q_{fin}}{q_{no\,fin}} = \frac{q_{fin}}{hA_b(T_b - T_\infty)} = \frac{\eta_{fin} h A_{fin}(T_b - T_\infty)}{h A_b(T_b - T_\infty)} = \frac{\eta_{fin} A_{fin}}{A_b}.$$

Therefore, the fin effectiveness can be determined easily when the fin efficiency is known, or vice versa.

The rate of heat transfer from a sufficiently long fin or uniform cross section under steady conditions is given by Equation. Substituting this relation into Equation, the effectiveness of such a long fin is determined to be,

$$\varepsilon_{fin} = \frac{q_{fin}}{q_{no\,fin}} = \frac{(T_b - T_\infty)}{h A_b(T_b - T_\infty)} = \sqrt{\frac{kP}{hA_c}}$$

Since A_c=$A_b$in this case. We can draw several important conclusions from the fin effectiveness relation above for consideration in the design and selection of the fins

- The thermal conductivity k of the fin material should be as high as possible. Thus it is no coincidence that fins are made from metals, with copper, aluminium, and iron being the most common ones. Perhaps the most widely used fins are made of aluminum because of its low cost and weight and its resistance to corrosion.
- The ratio of the perimeter to the cross sectional area of the fin P/A_c should be as high as possible. This criterion is satisfied by thin plate fins or slender pin fins.
- The use of fins is most effective in applications involving low convection heat transfer coefficient. Thus, the use of fins is more easily justified when the medium is a gas instead of a liquid and the heat transfer is by natural convection instead of by forced convection. Therefore, it is no coincidence that in liquid-to-gas heat exchangers such as the car radiator, fins are placed on the gas side.

When determining the rate of heat transfer from a finned surface, we must consider the unfinned portion of the surface as well as the fins. Therefore, the rate of heat transfer for a surface containing n fins can be expressed as,

$$q_{total,fin} = q_{unfin} + q_{fin} = hA_{unfin}(T_b - T_\infty) + h_{fin}\, hA_{fin}\,(T_b - T_\infty)$$

$$q_{total,fin} = (hA_{unfin} + h_{fin}\, hA_{fin})\,(T_b - T_\infty)$$

We can also define an overall effectiveness for a finned surface as the ratio of the total heat transfer from the finned surface to the heat transfer from the same surface if there were no fins,

$$\varepsilon_{fin} = \frac{q_{fin}}{q_{no\,fin}} = \frac{h\left(A_{unfin} + \eta_{fin} A_{fin}\right)\left(T_b - T_\infty\right)}{h\,A_{no\,fin}\left(T_b - T_\infty\right)}$$

Where A_{nofin} is the area of the surface when there are no fins, $A_{unfin}$is the total surface area of all the fins on the surface, and A unfin is the area of the unfinned portion of the surface. Note that the overall fin effectiveness depends on the fin density (number of fins per unit length) as well as the effectiveness of the individual fins. The overall effectiveness is a better measure of the performance of a finned surface than the effectiveness of the individual fins.

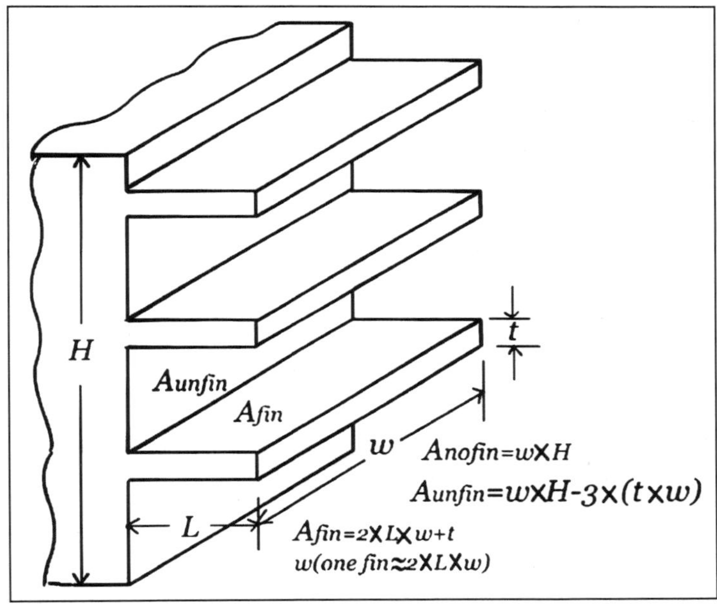

ONE DIMENSIONAL HEAT CONDUCTION EQUATION

There is an electrical analogy with conduction heat transfer that can be exploited in problem solving. The analog of $\dot{Q}$ is current, and the analog of the temperature difference, $T_1 - T_2$, is voltage difference. From this perspective the slab is a pure resistance to heat transfer and we can define,

$$\dot{Q} = \frac{T_1 - T_2}{R}$$

where $R = L/kA,$, the thermal resistance. The thermal resistance R increases as L increases, as A decreases, and as k decreases.

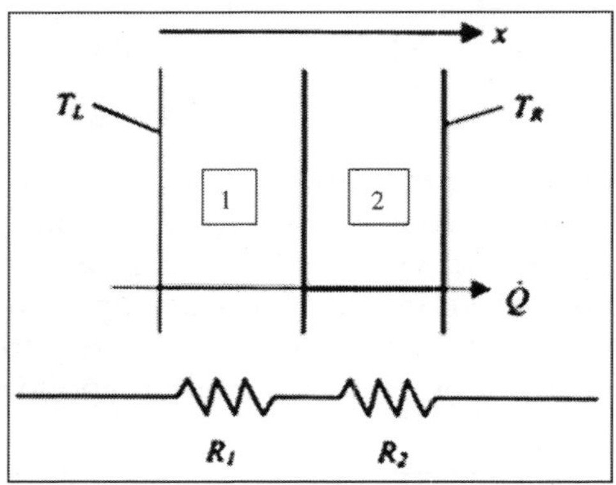

Fig. Heat Transfer Across a Composite Slab (series thermal resistance)

The concept of a thermal resistance circuit allows ready analysis of problems such as a composite slab (composite planar heat transfer surface). In the composite slab shown in Figure, the heat flux is constant with x. The resistances are in series and sum to $R = R_1 + R_2$. If T_L is the temperature at the left, and T_R is the temperature at the right, the heat transfer rate is given by,

$$\dot{Q} = \frac{T_L - T_R}{R} = \frac{T_L - T_R}{R_1 + R_2}.$$

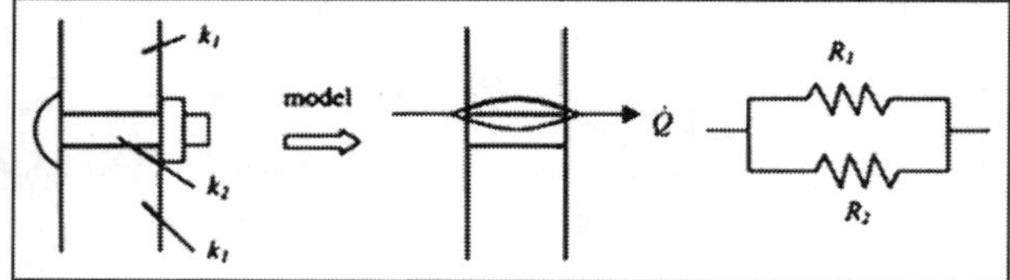

Fig. Heat Transfer for a Wall with Dissimilar Materials (Parallel Thermal Tesistance)

Another example is a wall with a dissimilar material such as a bolt in an insulating layer. In this case, the heat transfer resistances are in parallel. Figure shows the physical configuration, the heat transfer paths and the thermal resistance circuit.

For this situation, the total heat flow $\dot{Q}$ is made up of the heat flow in the two parallel paths, $\dot{Q} = \dot{Q}_1 + \dot{Q}_2$, with the total resistance given by,

$$\frac{1}{R} = \frac{1}{R_1} + \frac{1}{R_2}.$$

More complex configurations can also be examined; for example, a brick wall with insulation on both sides.

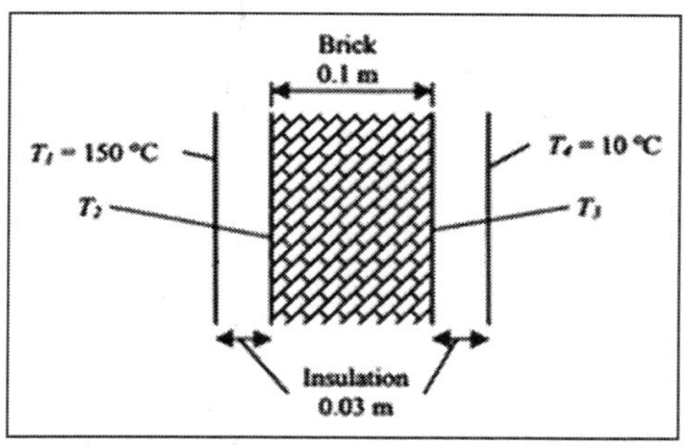

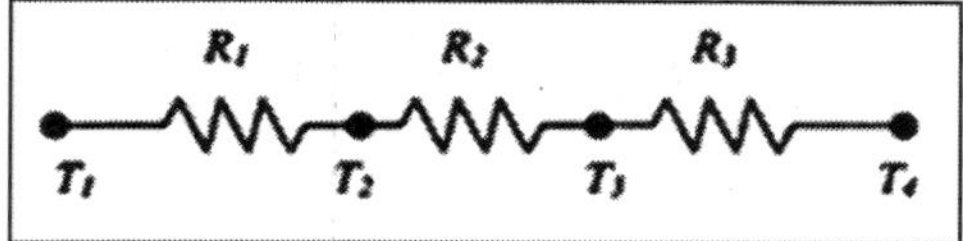

Fig. Heat Transfer through an Insulated Wall

The overall thermal resistance is given by,

$$R = R_1 + R_2 + R_3 = \frac{L_1}{k_1 A_1} + \frac{L_2}{k_2 A_2} + \frac{L_3}{k_3 A_3}.$$

Some representative values for the brick and insulation thermal conductivity are:

$$k_{brick} = k_2 \qquad = 0.7 W / m - K$$

$$k_{insulation} = k_1 = k_3 \quad = 0.07 W / m - K$$

Using these values, and noting that $A_1 = A_2 = A_3 = A$, we obtain,

$$AR_1 = AR_3 = \frac{L_1}{k_1} = \frac{0.03 \text{ m}}{0.07 \text{ W/m-K}} = 0.42 \text{ m}^2 \text{ K/W}$$

$$AR_2 = \frac{L_2}{k_2} = \frac{0.1 \text{ m}}{0.7 \text{ W/m-K}} = 0.14 \text{ m}^2 \text{ K/W}.$$

This is a series circuit so,

$$\dot{q} = \frac{\dot{Q}}{A} = \text{ constant throughout } = \frac{T_1 - T_4}{RA} = \frac{140 \text{ K}}{0.98 \text{ m}^2\text{K/W}} = 142 \text{ W/m}^2.$$

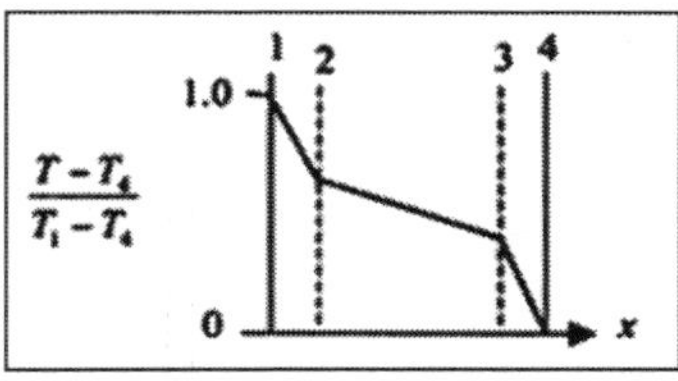

Fig. Temperature Distribution through an Insulated Wall

The temperature is continuous in the wall and the intermediate temperatures can be $\dot{Q}$ is constant across the slab. For example, to find T_2:

$$\dot{q} = \frac{T_1 - T_2}{R_1 A} = 142 \text{ W/m}^2.$$

This yields $T_1 - T_2 = 60K$ or $T_2 = 90°C$.

The same procedure gives $T_3 = 70°C$. As sketched in Figure, the larger drop is across the insulating layer even though the brick layer is much thicker.

TRANSIENT RESPONSE OF A STRAIGHT FIN

THE CIRCULAR FIN

Analytical expressions are developed to describe the transient response of a straight fin when changes occur in the temperature of the surrounding fluid. Initially, the fins may have the temperature of the surrounding fluid or they may be at some steady state. Results are presented for step and sinusoidal changes in the fluid temperature, but it is also shown how to simply extend the model so as to cover the cases when the fin is subjected to linear and higher order changes of fluid temperature. It was shown that the rate of heat removal is not enhanced by subjecting a fin to a periodically changing fluid temperature as opposed to a steady average temperature.

Problem

A pipe of radius R_0 has a circular fin of radius R_1 and thickness $2B$ on it (as shown in the figure below). The outside wall temperature of the pipe is T_w and the ambient air temperature is T_a. Neglect the heat loss from the edge of the fin (of thickness $2B$). Assume heat is transferred to the ambient air by surface convection with a constant heat transfer coefficient h.

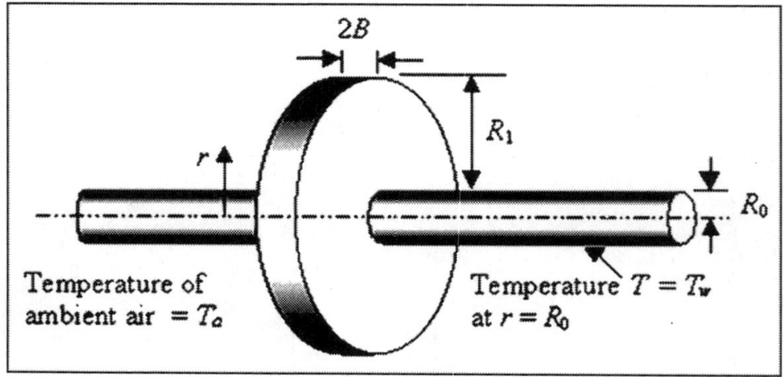

Fig. Radial Circular Fin on Heated Pipe.

- Starting with a shell thermal energy balance, derive the differential equation that describes the radial temperature distribution in the fin.
- Obtain the radial temperature distribution in the circular fin.
- Develop an expression for the total heat loss from the fin.

Solution

A circular fin is also called a radial fin or circumferential fin.

From a thermal energy balance over a thin cylindrical ring of width Dr in the circular fin, we get,

Rate of Heat In – Out + Generation = Accumulation

The accumulation term (at steady-state) and the generation term will be zero. So,

$$(2\pi r\, 2B\, q_r)|_r - (2\pi r\, 2B\, q_r)|_{r+\Delta r} - 2(2\pi r\Delta r)\, h\, (T - T_{\acute{a}}) = 0$$

where h is the (constant) heat transfer coefficient for surface convection to the ambient air and q_r is the heat flux for conduction in the radial direction.

Dividing by 4p B Δr and taking the limit as Δr tends to zero,

$$\lim_{\Delta\to 0} \frac{(rqr)|_{r+\Delta r} - (rqr)|_r}{\Delta r} = -\frac{h}{B} r(T - T_a)$$

$$\frac{d}{dr}(rqr) = -\frac{h}{B} r(T - T_a)$$

If the thermal conductivity k of the fin material is considered constant, on substituting Fourier's law ($q_r = -k\frac{dT}{dr}$) we get,

$$\frac{d}{dr}\left(r\frac{dT}{dr}\right) = \frac{h}{kB} r(T - T_a)$$

Let the dimensionless excess temperature be denoted by $q = (T - T_a)/(T_w - T_a)$. Then,

$$\frac{d}{dr}\left(r\frac{d\theta}{dr}\right) = \frac{h}{kB} r\theta$$

Using the chain rule on the left-hand side of the above equation and dividing throughout by r,

$$\frac{d^2\theta}{dr^2} = \frac{1}{r}\frac{d\theta}{dr} - \frac{h}{kB} - \theta = 0$$

Alternatively, starting with the general expression for the cooling fin, we have,

$$\frac{d}{dz}\left(kA\frac{d\theta}{dz}\right) = hP\theta$$

For the radial fin, the cross-sectional area (for conduction) is $A = 2pr2B$ and the perimeter (for surface convection) is $P = 4p\ r$. Also, $r = R_0 + z$, where z is the coordinate measuring distance from the outside wall of the pipe. Substituting these expressions for $A$and P with $dr = dz$, we get,

$$k\frac{d}{dr}\left(2\Pi r\ 2B\frac{d\theta}{dr}\right) = h4\Pi r\theta\ dr$$

The above equation on simplifying is identical to equation.

Equation above is a modified Bessel equation of order zero. Its solution is,

$$\theta = C_1\ I_0(cr) + C_2\ K_0(cr)$$

where $c^2 = h/(kB)$. Note that I_0 and K_0 are modified Bessel functions (of order zero) of first and second kind, respectively.

The integration constants C_1 and C_2 are determined using the boundary conditions:

BC 1: $r = R_0, \theta = 1 \Rightarrow 1 = C_1I_0(cR_0) + C_2K_0(cR_0)$

BC 2: $r = R_1, \dfrac{d\theta}{dr} = 0 \Rightarrow 0 = C_1cI_1(cR_1) - C_2cK_1(cR_1)$

The second boundary condition suggests no heat loss through the edge of the circular fin (of thickness $2B$), and requires the evaluation of derivatives of Bessel functions as given below:

$$\frac{d}{dr}I_0(cr) = \left[I_{-1}(cr)\right]c = cI_1(cr)$$

and

$$\frac{d}{dr}k_0(cr) = \left[K_{-1}(cr)\right]c = cK_1(cr)$$

Equations may be solved to yield C_1 and C_2. Thus,

$$C_1 = \frac{K_1(cR_1)}{I_0(cR_0)K_1(cR_1) + I_1(cR_1)K_0(cR_0)}\ and\ C_2\frac{I_1(cR_1)}{I_0(cR_0)K_0(cR_0)}$$

On substituting the integration constants, the dimensionless temperature profile is,

$$\theta = \frac{T - T_\alpha}{T_\omega - T_\alpha} = \frac{k_1(cR_1)I_0(cr) + I_1(cR_1)K_0(cr)}{K_1(cR_1)I_0(cR_0) + I_1(cR_1)K_0(cR_0)}$$

The heat flux at the base of the fin ($r = R_0$) is given by,

$$qr\big|_r = R_0 = \left[-k\frac{dT}{dr}\right]_{r=R_0} = -k(T_\omega - T_\alpha)\frac{K_1(cR_1)cI_1(cR_0) - I_1(cR_1)cK_1(cR_0)}{K_1(cR_1)I_0(cR_0) + I_1(cR_1)K_0(cR_0)}.$$

On multiplying the heat flux by the cross-sectional area for heat conduction, the total heat loss from the fin is obtained as,

Total heat loss from fin

$$= 4\pi R_0 Bk(T_\omega - T_\alpha)c\frac{k_1(cR_0)I_1(cR_1) - I_1(cR_0)K_1(cR_1)}{K_1(cR_1)I_0(cR_0) + I_1(cR_1)k_0(cR_0)}.$$

HEAT FLOW THROUGH RADIAL SYSTEMS

ORTHOGONAL-MODE ROTATION

In this entry, these effects are reviewed for the special case where the duct is constrained to rotate about an axis which is perpendicular to its axis of symmetry. This rotating configuration is referred to asorthogonal-mode rotation and is illustrated in Figure. Coriolis forces, in this case, act as a source term for the generation of a cross stream secondary flow as indicated in the figure. Thus a mechanism exists to cause relatively cool fluid from the central area of the tube to move towards the trailing edge. If the duct wall is heated this suggests that Coriolis forces tend to induce a relatively better cooled trailing edge region. With significant density variations, associated with heated flow, a centripetal buoyancy effect is produced analogous to the earth's gravitation field with a stationary vertical duct. Thus buoyancy effects will vary according to the direction of flow through the duct. In other words, buoyancy will change direction according to whether the flow is radially outward or inward. In the present article, only radially outward flow is considered for reasons of space.

DEVELOPED LAMINAR FLOW

Mori and Nakayama (1968) analysed laminar fully developed heat transfer with a uniformly heated tube of circular cross section using a momentum integral technique. These authors did not include buoyancy in their analysis and characterized Coriolis effects using a rotational form of the Reynolds Number, Jd ($\Omega d^2 \varrho/\eta$) where the velocity term involves the product of the tube diameter and the angular velocity. They did not allow for circumferential variations in heat transfer due to the Coriolis induced secondary flow owing to their choice of thermal boundary conditions, but recommendations for estimating a mean heat transfer coefficient over the circumference were proposed. Depending on the relative value of the rotational Reynolds number with respect to the through flow Reynolds number and the fluid Prandtl Number, two sets of recommendations were made.

Figure illustrates the typicalenhancement for a uniformly heated tube which resulted from the analysis. Here "enhancement" is defined as the ratio of the developed Nusselt Number obtained with rotation compared to the nonrotating value for a specified Reynolds number. The authors repeated the analysis for a constant wall temperature and found that the relative enhancement in heat transfer was insensitive to the thermal boundary condition.

The equations recommended are likely to give useful results for a mean circumferential heat transfer coefficient but will become less reliable if significant buoyancy effects are present since buoyancy was not included in the analysis. The equations derived do not permit the circumferential variation in heat transfer, due to the Coriolis driven secondary flow, to be evaluated. A number of experimental studies have been reported for laminar heat or mass transfer for this rotating geometry, prior to the early 1980's. Unfortunately the data is not conclusive, probably due to the difficulty of the experimental regime. A review of all data up to 1980 is given by Morris (1981).

DEVELOPING AND DEVELOPED TBRBULENT FLOW

Turbulent flow in orthogonally rotating ducts has been the subject of considerable theoretical and experimental study since about 1980 due to its importance for the internal cooling of gas turbine rotor blades. Morris and Ayhan (1979) first demonstrated experimentally that the combined influence of Coriolis forces and centripetal buoyancy could produce regions of impaired heat transfer, relative to the stationary duct case, on the leading edge of a circular tube. This was important from the design viewpoint in turbine blades since local hot spots might occur on the leading edge. Morris reviews virtually all published work prior to 1980 and a number of early results are available in this text.

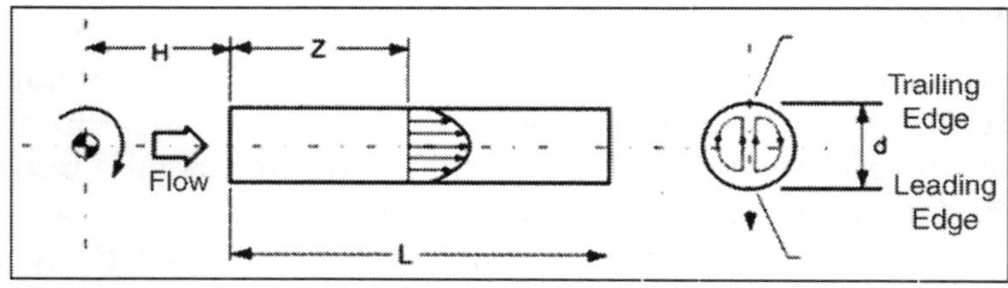

Fig. Orthogonal-mode Rotation.

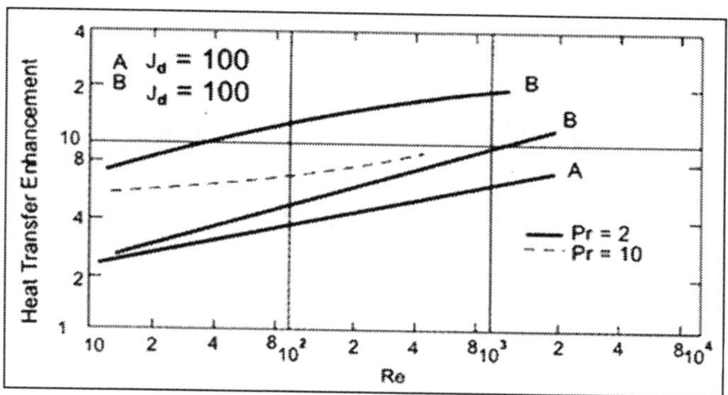

Fig. Fully Developed Laminar Flow.

An examination of the momentum and energy conservation equations for this rotating geometry suggests that the Nusselt number, evaluated at a specified location, z, in the duct, will be functionally related to the usual nondimensional groups which control forced convection (*i.e.* the through flow Reynolds number, Re, the Prandtl number of the fluid, Pr) together with two additional terms which arise from the Coriolis forces and the centripetal buoyancy. The nondimensional group which derives from the Coriolis term is the Rossby Number. Ro, which is the ratio of the coolant velocity along the duct to an angular speed term involving the product of the angular velocity with the tube diameter.

Thus,

$$Ro = w_m / \Omega d$$

The second additional term is a so-called buoyancy parameter, Bu, defined as,

$$Bu = (H + z)(T_{w,z} - T_{b,z}) / T_{b,z} d$$

where Tw,z and Tb,z refer to the local wall and fluid bulk temperatures, respectively.

We expect therefore that the local Nusselt number, NuR,z, at some specified position along the duct will have functional form,

$$Nu_{R,z} = \Psi\{Re, Pr, z/d, Ro, Bu\}$$

and the function, ψ, will be also dependent on the circumferential location at a give z-value.

Subsequent to 1980, a number of theoretical and experimental studies have been undertaken, see for example Wagner et al. Iacovides and Launder, Taylor et al. and Morris and Salemi have attempted to evaluate the appropriate functional form of the above equation. The references cited in these papers give a full coverage of the total literature currently available with radially outward, radially inward flow and also the effect of internal ribbing. The works of Wagner et al. and Morris and Salemi confirm that Coriolis forces enhance heat transfer on the trailing edge of smooth walled circular ducts relative to the case resulting from a similar value of through flow Reynolds number. Additionally, as the buoyancy parameter, Bu, is increased at fixed values of the Reynolds number and Rossby number, the heat transfer also increases. On the leading edge it is evident that rotation initially impairs heat transfer relative to the stationary duct case. However, at higher rotational speeds this reduction seems to be arrested with subsequent improvement of the relative heat transfer.

Morris and Chang (1994), in an as yet unpublished experimental study, have attempted to correlate their own new data for a smooth-walled circular tube and the data of Wagner et al. (1989) using a particular mathematical structure for ? based on physical reasoning concerning the combination of Coriolis and buoyancy interactions. They have proposed that the heat transfer ratio at different axial locations downstream in the duct may be expressed as,

$$Nu_{R,z} / Nu_{o,z} = 1 + \Phi\{(H + z)(T_{w,z} - T_{b,z}) / T_b dRo^2\}$$

where $Nu_{o,z}$ is the zero speed Nusselt number obtained at the same Reynolds number as the rotating case and F is some function. The argument of the function, F is a combined Coriolis/buoyancy parameter.

Figure shows the heat transfer ratio data produced by Morris and Chang plotted against the combined Coriolis/buoyancy parameter proposed. In this figure data from Wagner et al. is also compared at an axial location equivalent to 10.7 effective diameters downstream of entry to the duct. Although some data scatter is evident the proposed correlating function appears to have merit. Thus this figure may be used to estimate the level of heat transfer ratio expected at a particular operating condition.

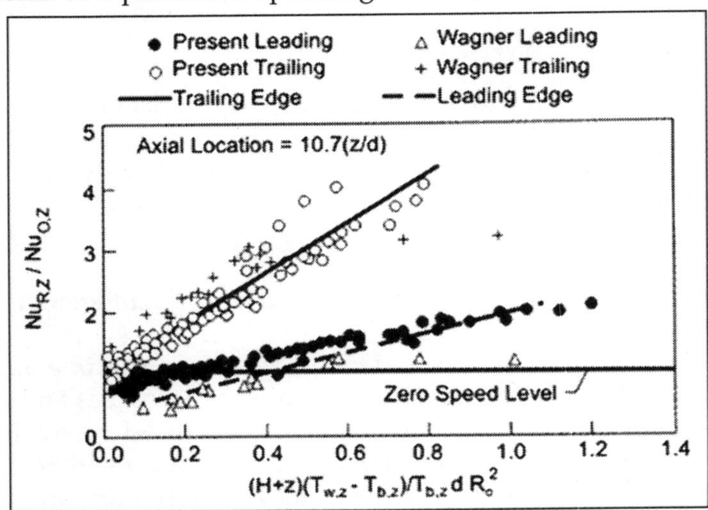

Fig. Effect of Rotation on Heat Transfer on the Leading and Trailing Edges.

HEAT FLOW THROUGH A PLANE WALL

Energy is transferred from more energetic to less energetic molecules when neighbouring molecules collide. Conductive heat flow occur in direction of the decreasing temperature since higher temperature are associated with higher molecular energy.

Fourier's Law express conductive heat transfer as,

$$q = k\ A\ dT / s$$

where,

A = heat transfer area (m^2, ft^2)

k = thermal conductivity of the material (W/m.K or W/m °C, Btu/(hr oF ft^2/ft))

dT = temperature difference across the material (K or °C, °F)

s = material thickness (m, ft)

HEAT TRANSFER BY CONDUCTION

A plane wall constructed of solid iron with thermal conductivity 70 W/ moC, thickness 50 mm and with surface area1 m by 1 m, temperature 150 °C on one side and 80oC on the other.

Conductive heat transfer can be calculated as:

q = (70 W/m °C) (1 m) (1 m) ((150 °C) – (80 °C)) / (0.05 m)
= 98,000 W
= 98 kW

HEAT TRANSFER THROUGH PLANE WALLS IN SERIES

Heat conducted through several walls in good thermal contact can be expressed as,

$$q =(T_1 - T_n) /((s_1/k_1A) + (s_2/k_2A) +... + (sn/knA))$$

where,

T1 = temperature inside surface (K or °C, °F)

TN = emperature inside surface (K or °C, °F)

A furnace wall of 1 m^2 consist of a 1.2 cm thick stainless steel inner layer covered with a 5 cm this outside insulation layer of asbestos board insulation. The inside surface temperature of the steel is 800 K and the outside surface temperature of the asbestos is 350 K. The thermal conductivity for stainless steel is 19 W/mK and for asbestos board 0.7 W/mK.

The conductive heat transport through the wall can be calculated as

q =((800 K) – (350 K)) /(((0.012 m)/(19 W/mK)(1 m^2)) + ((0.05 m)/(0.7 W/ mK)(1 m^2)))
= 6245 W

THERMAL CONDUCTIVITY AND COMMON UNITS

- Btu/(h ft^2 oF/ft)

- Btu/(h ft^2 oF/in)
- Btu/(s ft^2 oF/ft)
- MW/(m^2 K/m)
- kW/(m^2 K/m)
- W/(m^2 K/m)
- W/(m^2 K/cm)
- W/(cm^2 °C/cm)
- W/(in^2 °F/in)
- kJ/(h m^2 K/m)
- J/(s m^2 °C/m)
- kcal/(h m^2 °C/m)
- cal/(s cm^2 °C/cm)

HEAT TRANSFER IN LAMINAR FLOW IN TUBES

Laminar flow generally happens when dealing with small pipes and low flow velocities. Laminar flow can be regarded as a series of liquid cylinders in the pipe, where the innermost parts flow the fastest, and the cylinder touching the pipe isn't moving at all.

Shear stress depends almost only on the viscosity -μ- and is independent of density - ρ

SOLUTION FOR CONSTANT WALL TEMPERATURE

It is steady, constant property (including density), no mass transfer (single phase),, and laminar flow of Newtonian fluid. The flow is also axisymmetric (nothing depends on θ). Ignore body forces. Velocity (only) is fully developed

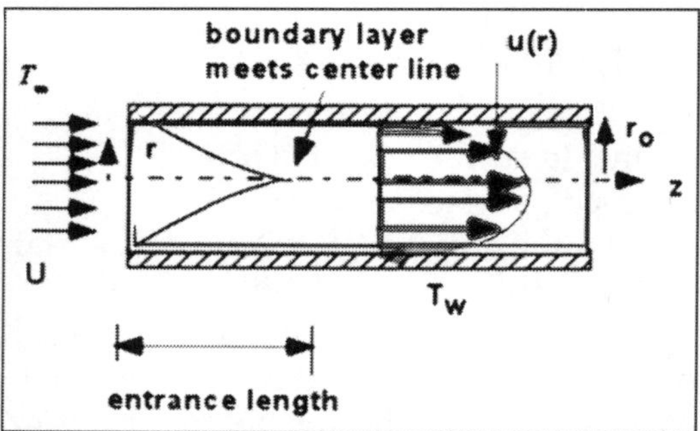

We will consider only the energy equation for further development.

The continuity and momentum equation yield the parabolic velocity profile:

$$V_L(r) = 2V_m\left(1 - \frac{r^2}{r_0^2}\right)$$

For constant wall temperature,

$$T(r_0, z) = T_\omega; \ \frac{\partial T}{\partial_r}(0, z) = 0$$

$$\frac{\partial T}{\partial z} = \tilde{T}\frac{dT_m}{dz} = \frac{T_w - T}{T_w = T_m}\frac{dT_m}{dz}$$

Energy Equation,

$$\left[\frac{1}{r}\frac{\partial}{\partial r}\left(r\frac{\partial T}{\partial r}\right)\right] = \frac{\rho C_p}{k}\left[V_z\frac{\partial T}{\partial_z}\right]$$

$$\frac{1}{r}\frac{\partial}{\partial r}\left(r\frac{\partial T(r,z)}{\partial r}\right) = \frac{2V_m}{\alpha}\left(1 - \frac{r^2}{r_0^2}\right)\left(\frac{T_w - T_{(r,z)}}{T_w - T_{m(z)}}\right)\frac{dT_m}{d_z}$$

$$T(r_0, z) = T_w; \frac{\partial T}{\partial r}(0, z) = 0$$

Series Solution,
One of the solution to the equation is a series solution by Bhatti

$$\tilde{T} = \sum_{n-0}^{\infty} C_{2n}\left(\frac{r}{r_0}\right)^{2n}$$

$$C_0 = 1; \ C_2 = -1.828397; \quad c_{2n} = \frac{2.704364^2}{(2n)^2}(C_{2n-4} - C_{2n-2})$$

Solution:
Nu = 3.657

DESCRIPTION

An alternate solution to the energy equation for pipe flow arises if we assume that the axial component $\partial T/\partial z$ of the temperature gradient is proportional to the local temperature difference $T_\omega - T$ between the wall and a fluid particle:

$$\frac{\partial T}{\partial z} = k(T_w - T)$$

where the proportionality constant κ has the same value throughout the flow field. Along the tube wall, where $T = T_w$, $\partial T_w/\partial_z$ is therefore zero and the wall temperature T_w is constant. Equation then may be integrated:

$$\frac{\partial(T_w - T)}{\partial z} = -\kappa(T_w - T)$$

$$T_w - T = (T_w - T)_{z=0} \exp(-\kappa z)$$

showing that the temperature difference between pipe wall and the fluid along a streamline decreases exponentially in the axial direction; *i.e.*, the fluid warms up to the wall temperature as it flows through the pipe.

With this assumption of constant wall temperature, we may combine equations to give the energy equation:

$$\alpha\left[\frac{1}{r}\frac{\partial}{\partial r}\left(r\frac{\partial T}{\partial r}\right)\right] = V_z\kappa(T_w - T) = \frac{\kappa\bar{V}}{2}\left(1 - \frac{r^2}{a^2}\right)(T_w - T)$$

The first integral of the energy equation, given, then has the form;

$$q = \frac{\kappa\rho\hat{c}_p}{2\pi a}\int_0^{\pi}(T_w - T)(2\pi r)$$

$$\frac{\kappa a}{2} = \frac{q}{p\hat{c}\bar{V}(T_w - \bar{T})}$$

$$= St = \frac{N_{uD}}{\Pr R\varepsilon_D}$$

$$\kappa D = 4\frac{Nu_D}{\Pr R\varepsilon_D}$$

where we have used equations. Thus we can determine the Nusselt number when we find the value of κa for the solution of the energy equation,

Using equation, the energy equation may be reduced to the dimensionless form:

$$\frac{1}{\varsigma}\frac{\partial}{\partial\varsigma}\left(\varsigma\frac{\partial}{\partial\varsigma}\left[\frac{T_w - T}{T_w - T_m}\right]\right) = -\frac{Nu_D}{2}\left(1 - \varsigma^2\right)\left[\frac{T_w - T}{T_w - T_m}\right]$$

$$\varsigma = \frac{r}{a}$$

with the boundary conditions that $T = T_w$ at $\varsigma = 1$ and $\partial T/\partial\varsigma = 0$ at $\varsigma = 0$. Equation must be solved numerically, and the boundary conditions can be met only by using a specific value for the Nusselt number:

$$Nu_D = 3.657; \quad T_w = const.$$

Equations link the axial temperature gradient and the Nusselt number for the case of constant wall temperature. In contrast to the case of constant heat transfer rate considered, both the mean temperature difference $T_w - \overline{T}$ and the heat transfer rate q decline exponentially with axially distance z, but their ratio and the Nusselt number remain constant.

The temperature distribution is given by the series:

$$\frac{T_w - T}{T_w - T_m} = \sum_{n=0}^{\infty} C_{2n} \left(\frac{r}{q}\right)^{2n}$$

$$C_0 = 1$$

$$C_2 = -\frac{Nu_D}{2}$$

$$C_{2n} = \frac{Nu_D}{2n^2}\left(C_{2n-4} - C_{2n-2}\right); \quad n \geq 2$$

About seven terms of this series are needed to give 1% accuracy.

The temperature distribution of equation for T_w *const*, expressed as a function of axial and radial distances, is plotted in figure. In contrast to the temperature field for q= *const* of figure the axial temperature gradient $\partial T / \partial z$ declines exponentially with z and the radial profile of temperature is more full.

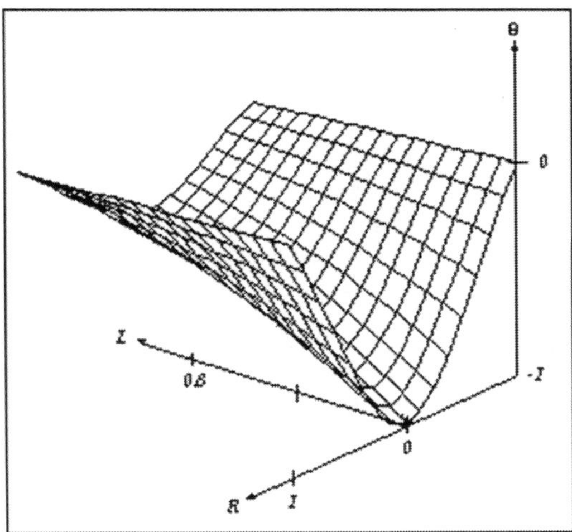

Fig. The Temperature Distribution $T\{r, z\}$ of a Fluid in Laminar Flow in a Pipe being Heated so as to Maintain a Constant Wall Temperature T_w, Expressed in terms of Dimensionless Coordinates:

$$\theta=\left(T\{r,z\}-T_w\{z\}\right)/\left(T_w\{z\}-T_m\{z\}\right),\ R\equiv 2r/D,\ z\equiv\left(4Nu_D/\mathrm{R}\varepsilon_D\,\mathrm{Pr}\right)\left(z/D\right).$$

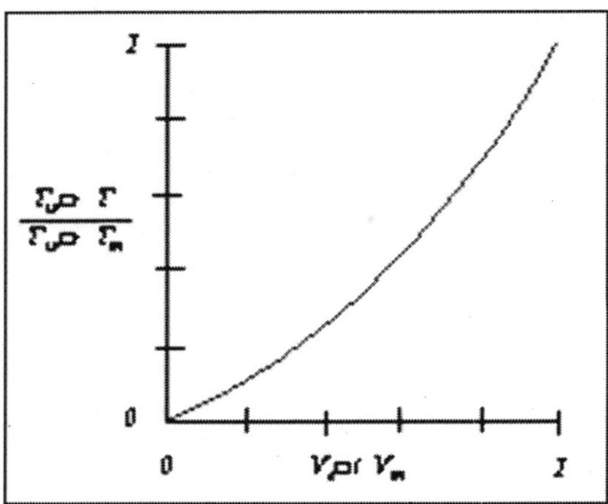

Fig. For Heat Transfer in a Pipe Flow having a Fixed Wall Temperature T_w, the Temperature Difference T_w-T varies more than Linearly with the Flow Velocity V_z. For this Case, Nu_D=3.657.

For fixed wall temperature, the variation of fluid temperature with velocity is plotted in figure. Compared with the constant heat transfer case shown in figure (b), the temperature profile is fuller and less linear, but not greatly different.

In equation we left out the term $\partial^2T/\partial z^2$, assuming it to be small compared with the left side. This is allowable if $\kappa a<<1$. Referring to equation, this constraint thus requires that $P_r\,\mathrm{R}\varepsilon_D>>1$. Most fluids have Prandtl numbers that are not much less than unity, so that the constraint is satisfied if $\mathrm{R}\varepsilon>>1$. However, liquid metals have very small Prandtl numbers and may require a more accurate analysis than that given above.

Example: In a laboratory experiment air at atmospheric pressure is to be heated from the ambient temperature $\mathbf{T_{in}}$=20C, by flowing it through a coil of copper tubing is immersed in a pan of boiling water that maintains the tube wall temperature T_w at 100C. Calculate the length L of tubing needed to ensure that the air outlet temperature T_{out} is not less than 99.9 C.

Soution: For air at atmospheric pressure and 20C, table gives v=1.51E(-5)m^2/s and Pr = 0.722. The air flow diameter Reynolds number is:

$$\mathrm{R}\varepsilon_D=\frac{\bar{V}D}{v}=\frac{(0.1m/s)(1E(-2)m)}{1.51E(-5)m^2/s}=64.52$$

which makes the fow laminar because $\mathrm{R}\varepsilon_D$< 2300. For heat transfer to a tube whose wall tempeature is constant, we evaluate at $z = L$ *to find:*

$$T_w = T_{out} = (T_w - T_{in})\exp(-\kappa L)$$

$$\kappa L = \ln\left(\frac{T_w - T_{in}}{T_w - T_{out}}\right) = \ln\left(\frac{80}{0.1}\right) = 8.987$$

But k is related to the Nusslet and Reynolds numbers by equation,

$$k = \frac{4\,Nu_D}{Dpr\,\mathrm{R}\,\varepsilon_D} = \frac{4(3.657)}{(0.01m)(0.722)(64.52)} = 31.40m^{-1}$$

where we have used the value of the Nusselt number given in equation for the constant wall temprature case. Thus the Required length L is:

$$L = \frac{8.987}{\kappa} = \frac{8.987}{31.40m^{-1}} = 0.286m$$

HEAT FLOW CONDITIONS OF THERMAL DIFFUSIVITY

Thermal diffusivity is an important property, required for such purposes under transient heat flow conditions, such as design applications, determination of safe operating temperature, process control, and quality assurance.

The flash method is used to measure values of thermal diffusivity, a, of a wide range of solid materials. It is particularly advantageous because of simple specimen geometry, small specimen size requirements, rapidity of measurement and ease of handling

Thermal diffusivity is a combination of physical properties (conductivity, density and specific heat capacity) (λ/∂cp) denoted by κ, naturally arising in the derivation of the conduction equation and having physical significance in the context of transient conduction processes. It may be regarded as the ratio of the Thermal Conductivity of a material to the specific heat capacity of that material. Thus copper, with a high value of eat capacity of that material. Thus copper, with a hi (= 11×10^{-5} $m^2 \bullet s^{-1}$), has a high conduction rate relative to its heat storage capacity and so responds quickly to changes in temperature.

Conversely insulating materials may have κ values 100 times smaller, these low κ values being a measure of the high thermal inertia of such materials. Values of κ are often calculated from the widely available data on λ, σ and cp but directly determined values are also available. Data values can be found in Touloukian and Ho (1977), an extensive data bank compiled by the Thermophysical Properties Research Center at Purdue University and continually extended and updated by CINDAS at the same university. Other readily available data books are those by Kaye and Laby (1986) and Perry (1984).

Extensive data banks are also available in computerized form and as CDROMs. Experimental methods for the determination of thermal diffusivity and conductivity of solids, liquids and gases are described by Tye (1969). κ values for gases show a strong dependence on temperature and pressure; methods for estimating ? values at temperature /pressure values outside tabulated ranges are described by Reid *et al.*

Thermal diffusivity of amorphous transparent conductive films, indium zinc oxide (IZO) films, with a thickness of 200 nm has been analysed quantitatively using a newly developed nanosecond thermoreflectance system. IZO films sandwiched by molybdenum (Mo) films were prepared on fused silica substrate by dc magnetron sputtering using an oxide ceramic IZO target (89.3 wt% In_2O_3 and 10.7 wt% ZnO). The resistivity, carrier density, and Hall mobility of the IZO films ranged from 4.2×10^{-4} to 22.7 Ωcm, from 2.6×10^{16} to 4.2×10^{20} cm^{-3}, and from 10 to 51 cm^2/V?s, respectively. The thermoreflectance signals were analysed based on an analytical solution of the one dimensional heat flow across the three-layered film (Mo/IZO/Mo) system.

The thermal diffusivity of the IZO films was $(0.6–1.3)\times10^{-6}$ m^2/s, depending on the electrical resistivity. The thermal conductivity carried by free electrons was estimated using the Wiedemann-Franz law. The phonon contribution to the heat transfer in IZO films with various resistivities was found to be almost constant (λ_{ph}= 1.85 W/m·K), which was about half of the one for polycrystalline indium tin oxide films.

INITIAL AND BOUNDARY CONDITIONS

Traditional computational fluid dynamics (CFD) methods solve the Navier–Stokes equations in order to directly determine the pressure and velocity (*i.e.*, macroscopic variables) in the fluid. In contrast, the Boltzmann equation governs a particle distribution function f(x,v,t), which depends on space x, particle velocity v, and time tsuch that,

$$\frac{\partial f}{\partial t}+\mathbf{v}\cdot\nabla f=Q(f,f),$$

where Q is a collision operator. It has been shown that in the limit of a small Knudsen number, the Navier–Stokes equations can be recovered from the Boltzmann equation if the Chapman–Enskog expansion is used.25 It is assumed that the solution of the Boltzmann equation can be expressed in the form of an infinite series,

$$f=f^{(0)}+f^{(1)}+f^{(2)}+\cdots,$$

where the lowest order term, f(0), is the equilibrium distribution function, corresponding to the Maxwell–Boltzmann distribution. If only the first two terms, f(0)and f(1), are taken into account in Eq., the Navier–Stokes equations are obtained. The collision operator Q in Eq. has a complex form, but it can be simplified by using the Bhatnagar–Gross–Krook (BGK) approximation,

$$Q(f,f) = -\frac{1}{\lambda}\left(f - f^{(0)}\right),$$

where λ is the relaxation time due to collision. The relationship between the macroscopic quantities (density ρ and momentum density ρu) and the distribution function f is given by,

$$\rho = \int f(\mathbf{x}, \mathbf{v}, t)\, d\mathbf{v}, \quad \rho\mathbf{u} = \int \mathbf{v} f(\mathbf{x}, \mathbf{v}, t)\, d\mathbf{v}.$$

The velocity space can be discretized into a finite set of velocities, vi, whose directions define a lattice. The corresponding distribution functions become fi(x,t)=f(x,vi,t). The corresponding discrete Boltzmann equation can be written as,

$$\frac{\partial f_i}{\partial t} + \mathbf{v}_i \cdot \nabla f_i = -\frac{1}{\lambda}\left(f_i - f_i^{(0)}\right).$$

In the present paper, a two-dimensional square lattice with nine-velocity directions is used. The discrete velocities are expressed in terms of the lattice directions, vi=cei, where c is the lattice speed, c=dx/dt. Here, dx and dt are the lattice space and time steps, respectively, and the lattice directions are indicated by,

$$\mathbf{e}_i = \begin{cases} (0,0), & i=0 \\ \left(\cos\left(\frac{\pi}{2}(i-1)\right), \sin\left(\frac{\pi}{2}(i-1)\right)\right), & i=1,2,3,4 \\ \sqrt{2}\left(\cos\left(\frac{\pi}{2}\left(i-\frac{9}{2}\right)\right), \sin\left(\frac{\pi}{2}\left(i-\frac{9}{2}\right)\right)\right), & i=5,6,7,8. \end{cases}$$

This model is commonly referred to as D2Q9. There are only three distinct values of speed in the lattice: 0 ("rest particle") and c and $c\sqrt{2}$ on the rectangular and oblique directions, respectively. The method deals with individualized particles that travel between two lattice nodes during one time step, such that it has a Lagrangian character.

THE HEAT TRANSFER OF THERMAL ENERGY

UNSTEADY STATE PROCESS

Heat transfer is the transfer of thermal energy from a body, at a high temperature, to another at a lower temperature. This transfer of thermal energy may occur under steady or unsteady state conditions. Under Steady state conditions the temperature within the system does not change with time. Conversely, under unsteady state conditions the temperature within the system does vary with time. Unsteady state conditions are a precursor to steady state conditions. No system exists initially under steady state conditions. Some time must pass, after heat transfer is initiated, before the system reaches steady state. During that period of transition the system is under unsteady state conditions.

Clearly, no system can remain under unsteady state conditions perpetually. The temperature of the system will eventually reach the temperature of the heat source, and once this happens, the system will be at steady state. Even if the amount of heat being transferred into the system is increased, at some point the system reaches its critical temperature and the energy transferred into it the starts causing phase changes within the system rather than temperatures increases. In food process engineering, heat transfer is very often in the unsteady state, in which temperatures are changing and materials are warming or cooling. Unfortunately, study of heat flow under these conditions is complicated. In fact, it is the subject for study in a substantial branch of applied mathematics, involving finding solutions for the Fourier equation written in terms of partial differentials in three dimensions. There are some cases that can be simplified and handled by elementary methods, and also charts have been prepared which can be used to obtain numerical solutions under some conditions of practical importance.

A simple case of unsteady state heat transfer arises from the heating or cooling of solid bodies made from good thermal conductors, for example a long cylinder, such as a meat sausage or a metal bar, being cooled in air. The rate at which heat is being transferred to the air from the surface of the cylinder is given by eqn.

$$q = \mathrm{d}Q/\mathrm{d}t = h_s A(T_s - T_a)$$

where T_a is the air temperature and T_s is the surface temperature.

Now, the heat being lost from the surface must be transferred to the surface from the interior of the cylinder by conduction. This heat transfer from the interior to the surface is difficult to determine but as an approximation, we can consider that all the heat is being transferred from the centre of the cylinder. In this instance, we evaluate the temperature drop required to produce the same rate of heat flow from the centre to the surface as passes from the surface to the air. This requires a greater temperature drop than the actual case in which much of the heat has in fact a shorter path.

Assuming that all the heat flows from the centre of the cylinder to the outside, we can write the conduction equation,

$$\mathrm{d}Q/\mathrm{d}t = (k/L)A(T_c - T_s)$$

where T_c is the temperature at the centre of the cylinder, k is the thermal conductivity of the material of the cylinder and L is the radius of the cylinder.

Equating these rates:

$$h_s A(T_s - T_a) = (k/L)A(T_c - T_s)$$

$$h_s(T_s - T_a) = (k/L)(T_c - T_s)$$

and so,

$$h_sL/k = (T_c - T_s)/ (T_s - T_a)$$

To take a practical case of a copper cylinder of 15 cm radius cooling in air $k_c = 380$ J m^{-1} s^{-1} °C^{-1}, $h_s = 30$ J m^{-2} s^{-1}°C^{-1}, $L = 0.15$ m,

$$(T_c - T_s)/ (T_s - T_a) = (30 \times 0.15)/380 = 0.012$$

In this case 99% of the temperature drop occurs between the air and the cylinder surface. By comparison with the temperature drop between the surface of the cylinder and the air, the temperature drop within the cylinder can be neglected.

On the other hand, if the cylinder were made of a poorer conductor as in the case of the sausage, or if it were very large in diameter, or if the surface heat-transfer coefficient were very much larger, the internal temperature drops could not be neglected.

This simple analysis shows the importance of the ratio:

$$\frac{\textit{Heat transfer coefficent at the surface}}{\textit{Heat conduc}\tan \textit{ce to the centre of the solid}} = h_S L / k$$

This dimensionless ratio is called the Biot number (Bi) and it is important when considering unsteady state heat flow. When (Bi) is small, and for practical purposes this may be taken as any value less than about 0.2, the interior of the solid and its surface may be considered to be all at one uniform temperature.

In the case in which (Bi) is less than 0.2, a simple analysis can be used, therefore, to predict the rate of cooling of a solid body.

Therefore for a cylinder of a good conductor, being cooled in air,

$$dQ = h_sA(T_s - T_a)\, dt$$

But this loss of heat cools the cylinder in accordance with the usual specific heat equation:

$$dQ = c\rho V dT$$

where c is the specific heat of the material of the cylinder, r is the density of this material and V is the volume of the cylinder. Since the heat passing through the surface must equal the heat lost from the cylinder, these two expressions for dQ can be equated:

$$c\rho V dT = h_sA(T_s - T_a)\, dt$$

Integrating between $T_s = T_1$ and $T_s = T_2$, the initial and final temperatures of the cylinder during the cooling period, t, we have:

$$- h_sAt/crV = \log_e (T_2 - T_a)/(T_1 - T_a)$$

$$\text{or } (T_2 - T_a)/(T_1 - T_a) = \exp(-h_sAt/crV))$$

For this case, the temperatures for any desired interval can be calculated, if the surface transfer coefficient and the other physical factors are known. This gives a reasonable approximation so long as (Bi) is less than about 0.2. Where (Bi) is greater than 0.2, the centre of the solid will cool more slowly than this equation suggests. The equation is not restricted to cylinders, it applies to solids of any shape so long as the restriction in (Bi), calculated for the smallest half-dimension, is obeyed.

Charts have been prepared which give the temperature relationships for solids of simple shapes under more general conditions of unsteady-state conduction.

These charts have been calculated from solutions of the conduction equation and they are plotted in terms of dimensionless groups so that their application is more general. The form of the solution is:

$$f\{(T - T_0)/(T_i - T_0)\} = F\{(kt/crL^2)(h_sL/k)\}$$

where f and F indicate functions of the terms following, T_i is the initial temperature of the solid, T_0 is the temperature of the cooling or heating medium, T is the temperature of the solid at time t, (kt/crL^2) is called theFourier number (Fo) (this includes the factor k/cr the thermal conductance divided by the volumetric heat capacity, which is called the thermal diffusivity) and (h_sL/k) is the Biot number.

A mathematical outcome that is very useful in these calculations connects results for two- and three-dimensional situations with results from one-dimensional situations. This states that the two- and three-dimensional values called $F(x,y)$ and $F(x,y,z)$ can be obtained from the individual one-dimensional results if these are $F(x)$, $F(y)$ and$F(z)$, by simple multiplication:

$$F(x,y) = F(x)F(y)$$

and

$$F(x,y,z) = F(x)F(y)F(z)$$

Using the above result, the solution for the cooling or heating of a brick is obtained from the product of three slab solutions. The solution for a cylinder of finite length, such as a can, is obtained from the product of the solution for an infinite cylinder, accounting for the sides of the can, and the solution for a slab, accounting for the ends of the can. Charts giving rates of unsteady-state heat transfer to the centre of a slab, a cylinder, or a sphere, are given in Fig. On one axis is plotted the fractional unaccomplished temperature change,

$$(T - T_0)/(T_i - T_0).$$

On the other axis is the Fourier number, which may be thought of in this connection as a time coordinate. The various curves are for different values of the reciprocal of the Biot number, k/hr for spheres and cylinders, k/hl for slabs.

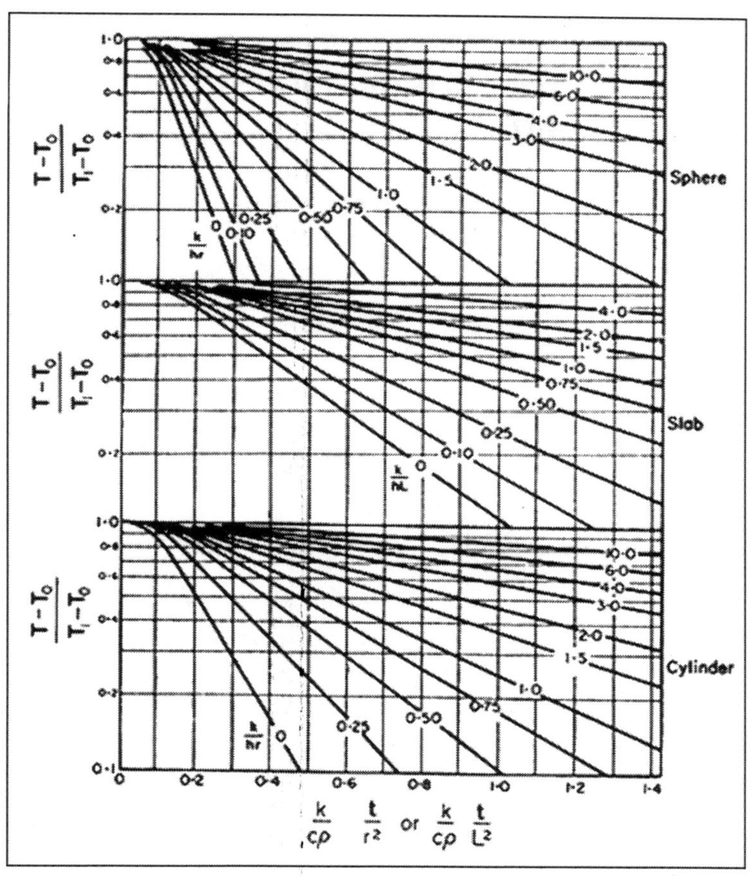

Fig. Transient Heat Conduction

THERMAL CONDUCTIVITY OF PERFECT INFINITE CRYSTALS

SOLIDS OF INFINITE THERMAL CONDUCTIVITY

The thermal conductivity of perfect infinite crystals of neon, argon, krypton, and xenon has been calculated numerically. It was assumed that the crystals possessed face centred cubic structure with the atoms interacting in pairs through a Lennard-Jones 12:6 potential energy function. The calculations considered only the effects of three-phonon interactions. It was possible to simplify the calculations by introducing "reduced" physical quantities. The thermal conductivity of each of the inert gas solids considered was obtained from the "reduced thermal conductivity" which was calculated for argon. Agreement with experimental data for neon, argon, and krypton was obtained for temperatures higher than those for which the effects of crystal size and lattice defects determine the thermal conductivity. This agreement suggests that for sufficiently high temperatures the thermal conductivity is determined by the effects of three-phonon Umklapp processes.

ANALYSIS OF HEAT CAPACITY

Lumped system analysis is the simplest and most convenient method that can be used to solve transient conduction problems. Hence, it is important to determine under what conditions it may be used with reasonable accuracy. The first step in establishing a criterion for the applicability of the lumped system analysis is to define a characteristic length as,

$$L_C = \frac{V}{A}$$

and a Biot number *Bi* as,

$$B_i = \frac{hL_C}{k}.$$

It can also be expressed as,

$$B_i = \frac{h}{k/L_C}\frac{\Delta T}{\Delta T} = \frac{\textit{Convection at the surface of the body}}{\textit{Conduction within the body}}.$$

When a solid body is being heated by the hotter fluid surrounding it (such as a potato being baked in a oven), heat is first convected to the body and subsequently conducted within the body. The Biot number is the ratio of the internal resistance of a body to heat conduction to its external resistance to heat convection. Therefore, a small Biot number represents small resistance to heat conduction, and thus small temperature gradients within the body.

Lumped system analysis assumes a uniform temperature distribution throughout the body, which will be the case only when the thermal resistance of the body to heat conduction (the conduction resistance) is zero. Thus, lumped system analysis is exact when $Bi=0$ and approximate when $Bi > 0$. Of course, the smaller the Biot number, the more accurate the lumped system analysis. Then the question we must answer is, how much accuracy are we willing to sacrifice for the convenience of the lumped system analysis?

Before answering this question, we should mention that a 20% uncertainty in the convection heat transfer coefficient *h* in most cases is considered "normal" and "expected." Assuming *h* to be constant and uniform is also an approximation of questionable validity, especially for irregular geometries. Therefore, in the absence of sufficient experimental data for the specific geometry under consideration, we cannot claim our results to be better than ± 20%, even when $Bi = 0$. This being the case, introducing another source of uncertainty in the problem will hardly have any effect on the overall uncertainty, provided that is minor. It is generally accepted that lumped system analysis is applicable if

$$B_i \leq 0.1$$

when this criterion is satisfied, the temperatures within the body relative to the surroundings (*i.e.*, $T - T_\infty$) remain within 5% of each other even for well-rounded geometries such as a spherical ball. Thus, when $Bi<0.1$, the variation of temperature with location within the body will be slight and can reasonably be approximated as being uniform.

The first step in the application of lumped system analysis is the calculation of the Biot number, and the assessment of the applicability of this approach. One may still wish to use lumped system analysis even when the criterion $Bi < 0.1$ is not satisfied, if high accuracy is not a major concern.

Note that the Biot number is the ratio of the convection at the surface to conduction within the body, and this number should be as small as possible for lumped system analysis to be applicable. Therefore, small bodieswith high thermal conductiviy are good candidates for lumped system analysis, especially when they are in medium that is a poor conductor of heat (such as air or another gas) and motionless. Thus, the hot small copper ball placed in quiescent air, discussed earlier, is most likely to satisfy the criterion for lumped system analysis. Consider heat transfer from a hot body to its cooler surroundings. Heat will be transferred from the body to the surrounding fluid as a result of a temperature difference. But this energy will come from the region near the surface, and thus the temperature of the body near the surface will drop. This creates a temperature gradientbetween the inner and outer regions of the body and initiates heat flow by conduction from the interior of the body towards the outer surface.

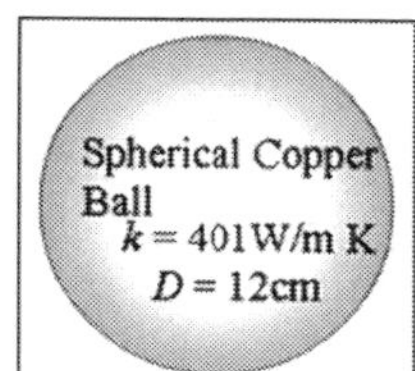

$$L_C = \frac{V}{A} = \frac{1/6\pi D^3}{\pi D^2} = 0.02m$$

$$Bi = \frac{h L_C}{k} = \frac{15 \times 0.02}{401} = 0.0075 < 0.1$$

When the convection heat transfer coefficient h and thus convection heat transfer from the body are high, the temperature of the body near the surface will drop quickly. This will create a larger temperature difference between the inner and outer regions unless the body is able to transfer heat from the inner and outer regions as fast. Thus, the magnitude of the maximum temperature difference within the body strongly depends strongly on the ability of a body to conduct heat towards its surface relative to the ability of the surrounding medium to convect this heat away from the surface. The Biot number is a measure of the relative magnitudes of these two competing effects.

Recall that the heat conduction in a specified direction n per unit surface area is expressed as,

$$\dot{q} = -k\frac{\partial T}{\partial n},$$

where,

$$\frac{\partial T}{\partial n}$$

is the temperature gradient and k is the thermal conductivity of the solid. Thus, the temperature distribution in the body will be *uniform* only when its thermal conductivity is infinite, and no such material is known to exist. Therefore, temperature gradients and thus temperature differences must exist within the body, no matter how small, in order for heat conduction to take place. Of course, the temperature gradient and the thermal conductivity are inversely proportional for a given heat flux. Therefore, the larger the thermal conductivity, the smaller the temperature gradient.

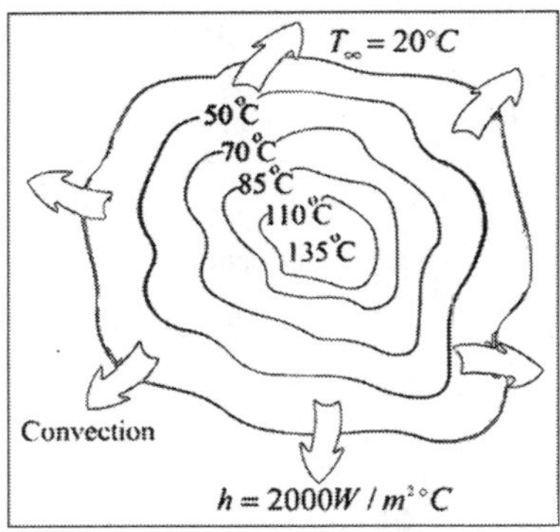

SOLUTION OF UNSTEADY STATE

A process is under consideration in which large cylindrical meat sausages are to be processed in an autoclave. The sausage may be taken as thermally equivalent to a cylinder 30 cm long and 10 cm in diameter. If the sausages are initially at a temperature of 21°C and the temperature in the autoclave is maintained at 116°C, estimate the temperature of the sausage at its centre 2 h after it has been placed in the autoclave. Assume that the thermal conductivity of the sausage is 0.48 J m^{-1} s^{-1} $°C^{-1}$, that its specific gravity is 1.07, and its specific heat is 3350 J kg^{-1} $°C^{-1}$. The surface heat-transfer coefficient in the autoclave to the surface of the sausage is 1200 J m^{-2} s^{-1} $°C^{-1}$.

This problem can be solved by combining the unsteady-state solutions for a cylinder with those for a slab, working from Fig.

(a) For the cylinder, of radius r = 5 cm (instead of L in this case)

$$\text{Bi} = h_s r/k = (1200 \times 0.05)/0.48 = 125$$

(Often in these systems the length dimension used as parameter in the charts is the half-thickness, or the radius, but this has to be checked on the graphs used.)
So 1/(Bi) = 8 × 10^{-3}
After 2 hours t = 7200 s
Therefore Fo = kt/crr^2 = (0.48 × 7200)/[3350 × 1.07 × 1000 × $(0.05)^2$] = 0.39 and so from Fig for the cylinder:

$$(T - T_0)/(T_i - T_0) = 0.175 = \text{say, } F(x).$$

(b) For the slab the half-thickness 30/2 cm = 0.15 m and so
Bi = $h_s L/k$ = (1200× 0.15)/0.48 = 3751/ Bi = 2.7 x 10^{-3}
t = 7200 s as before and kt/crL^2 = (0.48 × 7200)/[3350 × 1.07 × 1000 × $(0.15)^2$]
= 4.3 x 10^{-2} and so from Fig. for the slab:
$(T - T_0)/(T_i - T_0)$ = 0.98 = say, $F(y)$ So overall $(T - T_0)/(T_i - T_0) = F(x)$ $F(y)$ = 0.175 × 0.98 = 0.172
Therefore

$$\frac{T_2 - 116}{21 - 16} = 0.172$$

Therefore T_2 = 100°C.

INTERNAL FLOW - TURBULENT FLOW CONDITIONS

An example of internal flow is fluid flowing in a pipe. In pipes, turbulent flow has been traditionally assumed to occur at values of theReynolds number of about 2300. But that value depends on the "smoothness" of the pipe. For very smooth pipes, laminar flow can persist to Reynolds numbers of 10000 or more. The fluid velocity is zero at the wall and maximum at the center line. See the section entitled The Momentum Boundary Layer and Friction Factor for more details. Since the Prandtl number can vary widely depending of the fluid, much effort has been expended to develop appropriate equations as a function of Prantdl number. Whether the surface is at constant temperature or constant heat rate can affect the relationship among the Nusselt, Reynolds, and Prandtl numbers especially when the Prandtl number is less than 1. The following table gives some reasonable estimates of the relationship.

Prandtl Number	Heat Transfer Condition	Correlation
Pr < 0.1 (e.g. liquid metals)	Constant Heat Rate	
Pr < 0.1 (e.g. liquid metals)	Constant Temperature	
0.5 < Pr < 1 (e.g. gases)	Constant Heat Rate	
0.5 < Pr < 1 (e.g. gases)	Constant Temperature	
1.0 < Pr < ~20 (e.g. water and light liquids)	All	
Pr > ~20 (e.g. oils and viscous liquids)	All	

Surface roughness can affect heat transfer in the same way it can affect hydrodynamics. The section of the tutorial in the rotating cylinder electrode entitled Effect of Surface Roughness provides a graphic example of the effect of roughness on mass transfer. One way to estimate the Nusselt number for a rough surface is,

$$\frac{Nu_{rough}}{Nu_{smooth}} = \left(\frac{f_{rough}}{f_{smooth}}\right)^{1/h}$$

where Nu is the Nusselt number and f is the friction factor estimated for the smooth and rough pipe. Values of "n" of 0.5 and 1 have been reported. Data for gases seem to conform best to a value of "n" of 0.5. One complication is that estimation of the friction factor for the rough surface requires knowledge of the roughness factor, a quantity difficult to estimate under field conditions. This factor reflects the size of the protrusions above the surface. Usually, the greater the roughness factor, the greater the heat transfer through the surface to or from a fluid flowing over that surface.

EXTERNAL FLOW

An example of external flow is fluid flowing along a flat surface in which the boundary layer is small relative to the distance to any other surface. The flow region is bounded by a solid surface on one side and viscous flow on the other. Though in theory the fluid velocity increases from 0 at the wall to a free stream value at infinity, in reality it reaches approximately the free stream velocity relatively close to the wall. As discussed in the section on The Momentum Boundary Layer and Friction Factor laminar flow can be present up to large Reynolds numbers based on distance from the leading edge.

THERMAL BOUNDRY LAYER

The thickest layer which is subcritical to convection. Let the thickness of this layer be δd and the temperature difference across it be δT. Because convection transfers heat so efficiently with a small temperature gradient, almost the entire temperature gradient is present at the top and bottom thermal boundary layers.

When the surface temperature of a liquid is a harmonic function of time with a frequency ù, a temperature wave propagates into the liquid. The amplitude of this wave decreases exponentially with distance from the surface. The temperature oscillation is essentially concentrated in a layer of the order of $(2X/\omega)^{1/2}$, where x is the thermal conductivity of the liquid (thermal boundary layer). Depending on the phase, at certain positions below the surface the temperature gradient is directed downwards and if its magnitude is sufficiently large (the magnitude is a function of the amplitude and frequency of the surface oscillations) the liquid can become unstable with respect to the onset of convection.

In that case the convective motion may spread beyond the initial unstable layer. For low frequencies the stability condition can be derived from the usual static Rayleigh criterion, on the basis of the Rayleigh number and the average temperature gradient of the unstable layer. This quasi-static approach, used by Sal'nikov, is appropriate to those cases in which the period of the temperature oscillations is much larger than the characteristic time of the perturbations. But when these times are of the same order, the problem must be analysed in dynamic terms. The stability problem must then be formulated as a problem of parametricresonance excitation of velocity oscillations due to the action of a variable parameter-the temperature gradient. In an earlier work. we considered the problem of the stability of a horizontal layer of liquid with a periodically varying temperature gradient.

It was assumed that the thickness of the layer was much smaller than the penetration depth of the thermal wave, so that the temperature gradient could be assumed to be independent of position. In the present work we consider the opposite case, in which the liquid layer is assumed to be much larger than the penetration depth, i. e., a thermal boundary layer can be defined. The temperature gradient at equilibrium, which is a parameter in the equations determining the onset of perturbations, is here a periodic function of time and a relatively complicated function of the depth coordinate z.

DETERMINATION OF DIMENSIONLESS CORRELATION FOR FORCED CONVECTION

Forced convection heat transfer coefficients can be conveniently calculated with Excel spreadsheets. This type of calculation is typically based on a correlation of dimensionless numbers, usually Nusselt number in terms of Reynolds number and Prandtl number. Forced convection occurs with a fluid moving past a solid surface when the fluid and the solid are at different temperatures. Newton's Law of Cooling $[Q = hA(T_s - T_f)]$ is a simple expression for the rate for convective heat transfer.

The parameters in Newton's Law of Cooling are:

- Q is the rate of forced convection heat transfer (Btu/hr – U.S. or W – S.I.)

- T_s is the solid temperature (°F – U.S. or °C – S.I.)
- T_f is the fluid temperature (°F – U.S. or °C – S.I.)
- A is the area of the surface that is in contact with the fluid (ft^2 – U.S. or m^2 – S.I.)
- h is the convective heat transfer coefficient (Btu/hr-ft^2–°F – U.S. or W/m^2–K – S.I.)

DIMENSIONLESS NUMBERS – NUSSELT, REYNOLDS AND PRANDTL

Determining a good estimate for the heat transfer coefficient, h, is often the most difficult part of forced convection heat transfer calculations. The process for estimating the heat transfer coefficient for a particular forced convection application is often through a correlation for Nusselt number (Nu) in terms of Reynolds number (Re) and Prandtl number (Pr). These three dimensionless numbers are defined in the box below, along with the definitions of the parameters that appear in them.

$$Nu = \frac{h\,D}{k} \qquad Re = \frac{D\,V\rho}{\mu}$$

$$Pr = \frac{\nu}{\alpha} = \frac{\mu\,C_p}{k}$$

Dimensionless Numbers used in Forced Convection Heat Transfer Coefficient Correlations

D = characteristic length (diam. for pipe), ft or m
V = characteristic fluid velocity, ft/sec or m/s
k = thermal conductivity of the fluid, Btu/hr-ft-°F or kJ/hr-m-K
ρ = density of the fluid, slugs/ft^3 or kg/m^3
μ = viscosity of the fluid, lb-sec/ft^2 or N-s/m^2
h = heat transfer coefficient, Btu/hr-ft^2-°F or kJ/hr-m^2-K

NUSSELT NUMBER CORRELATIONS FOR TURBULENT FLOW INSIDE A PIPE

The Dittus Boelter equation, which has been around since 1930 (ref #1) has two forms as follows:

$Nu_o = 0.023\ Re^{0.8}Pr^{0.4}$, for 'heating' (temperature of wall > temperature of fluid), and

$Nu_o = 0.026\ Re^{0.8}Pr^{0.3}$, for 'cooling' (temperature of wall < temperature of fluid). Subject to: $0.7 < Pr < 120$; $10{,}000 < Re < 160{,}000$; $L/D > 10$ ($L/D > 50$ according to some authors). It is a rather simple equation to use, but has a fairly narrow range of acceptable values for Re and Pr.

$$Nu_0 = \frac{\left(\frac{f}{8}\right)(Re - 1000)Pr}{1 + 12.7\left(\frac{f}{8}\right)^{0.5}\left(Pr^{\frac{2}{3}} - 1\right)}$$

where,

$$f = (0.790 \ln \mathrm{Re} - 1.64)^{-2}$$

For:

$$0.5 < \mathrm{Pr} < 2000$$

0

$$3000 < \mathrm{Re} < 5 \times 10^6$$

Another correlation is shown in the box at the right. The range of values for Re and Pr for this correlation are also shown. This correlation can be used for a wider range of values of Re and Pr.

A third correlation is shown in the box at the left below. This correlation, described by Pethukov is only a minor variation of the second correlation shown at the right. This third correlation works for an even wider range of values for Re and Pr.

Index

A

Absolute Temperature 100, 209, 222
Adiabatic Expansion 215, 216
Adiabatic Process 1, 127, 129
Angular Momentum 226

B

Boltzmann 108
Bulk Liquid 234

C

Carnot Engine 215
Coefficients of Expansion 21
Conservation of Energy 101
Constant Pressure 137, 235, 236, 237

D

Definition of Gas 233

E

Enthalpy 233, 235, 236, 237
Enthalpy and Specific 233
Entropy and Cosmology 132
Entropy and Information Theory 130
Entropy Changes 138
Entropy in Astrophysics 128
Entropy in Quantum Mechanics 128
Equation of State 174, 230, 240, 241, 242, 243
Equilibrium 3, 4, 13, 19, 95, 96, 99, 100, 102, 107, 109, 110, 111, 116, 129, 130, 131, 132, 173, 174, 208, 217, 218
Equipartition Theorem 208, 209, 218, 225, 226

F

Fahrenheit 12, 13, 14, 19, 24
Feeling and Seeing 9
First Thermometer 11

G

Gas Law 235, 236, 238
Gas Pressure 22

H

Harmonic Oscillators 217
Heat Capacity 20, 128, 208, 209, 210, 218, 219, 220, 223, 224, 225, 226, 228, 229, 241
Heat Engines 210
Hypothesis 20, 23, 132

I

Ideal Gas 22, 128, 129, 174, 208, 215, 230, 231, 232, 240, 241, 242
Internal Energy 138, 236, 237
Isotherm 214, 215

L

Latent Heat 20
Law of Thermodynamics 8, 100, 101, 102, 128, 129, 130, 131, 138, 208, 210, 211, 212, 213, 214, 216, 219, 230
Liquid Nitrogen 234
Liquid Oxygen 234

M

Mean Values 4
Moment of Inertia 226, 227, 228

P

Perpetual Motion 212
Potential Energy 217, 218, 225, 227

Q

Quantum Theory 129
Quasi-static Processes 2

R

Reversible Process 138
Rotational Kinetic Energy 226

S

Specific Heats 19, 208, 220, 225, 227, 228, 229, 231, 232
Statistical Thermodynamics 129, 175, 208, 211

T

Thermodynamic Instruments 173
Thermodynamic Systems 172
Triple Point 240
Turbines 107
Types of Gases 233

U

Universal Gas 235, 237

Z

Zeroth Law 13, 110